Energy, Planning and Urban Form

Energy, Planning and Urban Form

Susan Owens

p **Pion Limited, 207 Brondesbury Park, London NW2 5JN**

© 1986 Pion Limited

All rights reserved. No part of this book may be reproduced in any form by photostat microfilm or any other means without written permission from the publishers.

ISBN 0 85086 118 7

Soc
HD
9502
A2
O98
1986

Printed in Great Britain by Page Bros (Norwich) Limited

Preface

The economic, social, and environmental benefits of using energy efficiently are now widely recognised, and since the 1973-1974 oil crisis there has been growing interest in ways in which land-use planning might contribute towards this objective in the longer term. The crisis emphasised how energy-profligate our built environment had become. It stimulated the search for a better understanding of the influence of energy systems on spatial structure, and of ways in which land use and built form affect levels of energy demand, especially for transport and space heating. Since new infrastructure is durable and the energy future is uncertain, it has become irresponsible to ignore the energy implications of alternative planning policies and to omit 'energy flexibility' from the list of land-use planning objectives.

As with most new fields of research, an initial flush of enthusiasm has subsequently been tempered by a more sober appraisal of reality. In much of the earlier work on energy and urban form too much significance was attached to energy costs and too little to other economic factors, politics, and social change. It has been justly criticised for falling into the trap of 'energy determinism'. Initially too there was insufficient attention to the many constraints which would inhibit the integration of energy considerations into the planning process. In this book, energy is treated not as a deterministic factor in urban change, but as an important permissive one; and as much emphasis is placed on constraints in formulating and implementing relevant policies, real problems from the point of view of those in planning practice, as on more theoretical aspects of identifying energy-efficient spatial structures. Experience of energy-integrated planning in the United States of America and in Denmark (discussed in the final chapter) shows that constraints need not preclude achievements when the political will exists.

The book is aimed at academics and practitioners. I hope that it begins to bridge the gulf between theory and practice which has often seemed unacceptably wide in a field with such important policy implications.

Susan Owens, Newnham College, Cambridge
March 1986

Acknowledgements

Many people, too numerous to mention individually, have given me information, inspiration, and moral support while I have been writing this book. I am grateful to all the planning officers who took trouble to complete a questionnaire and to write to me on the subject of energy and planning. A special word of thanks must go to Pion's anonymous referee, whose constructive comments helped to turn my first draft into something which is, I hope, more readable. I am also indebted to Ian Gulley, Arthur Shelley, and Mike Young, who drew the figures, and to Maria Constantinou, Jean Lucas, and Julie Fisher, who patiently typed and retyped the manuscript.

I should also like to thank the following for permission to reproduce copyright material.

W S Atkins and Partners, Surrey: table 4.3
Building Research Establishment, Watford: figure 4.1
Bureau of Planning, Portland, OR: table 6.1
Cambridge University Press, Cambridge: figures 3.1(b), 3.3
Conservation of Urban Energy Group, Melbourne, Victoria: figure 6.7
Controller of HMSO, London: table 4.3
J L Edwards, Department of Civil and Mineral Engineering, Minneapolis, MN: figure 3.1(c)
Energy Policy Butterworth Scientific, Guildford, Surrey: table 4.1
W H Freeman, New York: figure 5.1
P B Goodwin, Transport Studies Unit, Oxford: figure 2.1
G Hemmens, College of Architecture, Art, and Urban Planning, Chicago, IL: figure 3.1(a)
Institute of British Geographers, London: figure 3.5
International Institute for Applied Systems Analysis, Laxenburg: figures 1.2(a), 4.3
C Jensen-Butler, University of Århus, Århus County: table 6.5, figure 6.6
D Keplinger, Resources, Applications, Designs and Controls, Inc., Alexandra, VA: table 4.2
G Leach, IIED Earthscan, London: table 1.1
H Lee, Harvard University, Cambridge, MA: figure 6.1
H Mathieu, Secrétariat du Plan Urbain, Paris: figure 5.2
Milton Keynes Development Corporation, Milton Keynes: figure 6.3
P R Odell, on behalf of EURICES, Erasmus Universiteit, Rotterdam: figure 1.2(b)
P Rickaby, Open University, Milton Keynes: figures 5.4, 6.4, 6.5
W R D Sewell, University of Victoria, British Columbia: tables 6.2, 6.3
D Turrent for Energy Conscious Design, London: figures 4.2, 6.2
L S Windheim, Leo A Daly Co., San Francisco, CA: figure 4.4
L J Wood, University of Tasmania, Tasmania: table 2.1

For my parents

Contents

1 Energy, spatial structure, and planning
Introduction — 1
Energy and spatial structure — 2
The energy outlook in the 1980s — 5
Energy and planning — 8

2 Spatial responses to energy constraints
Introduction — 10
Reversal of trends? — 10
Response to energy constraints: some important issues — 11
   The nature of constraint — 11
   Inertia in the built environment — 13
   Identifying 'causes' — 13
   Nonspatial options for energy saving — 13
Energy constraints, behaviour, and spatial structures — 14
   Response to fuel price increases in transport — 15
   Physical shortfall — 17
   Summary — 18
Energy constraints and urban models — 18
Residential energy consumption and spatial structure — 21
Concluding comments — 23

3 Reducing transport energy requirements
Introduction — 24
   Approaches to defining energy-efficient structure — 24
Reducing transport energy requirements — 28
   The regional and subregional scales — 28
   The intraurban scale — 32
Summary so far — 35
Promoting energy-efficient transport — 36

4 Reducing energy requirements in buildings
Introduction — 41
Energy efficiency and built form — 41
Renewable energy and spatial structure — 44
   The local scale — 44
   Wider spatial implications of 'soft energy paths' — 49
Combined heat and power generation — 51
   Spatial requirements for CHP/DH — 52
   CHP/DH and the inner city — 57
   Solar powered DH? — 58
Conclusion — 58

5 Energy-efficient environments: synthesis and policy implications
Introduction 60
Energy-efficient environments—a synthesis 60
   Efficient characteristics 60
   Efficient structures 61
The potential for energy conservation 67
   The magnitude of savings 67
   Comparison with other conservation measures 69
Energy and planning: the policy implications 70
Constraints 74
   Economic constraints 74
   Social and environmental considerations 76
   Political constraints 78
Concluding comments 80

6 Energy-integrated planning in practice
Introduction 81
The USA: Davis, California, and Portland, Oregon 84
   Davis 84
   Portland 87
The United Kingdom 91
   Strategic planning 91
   Greater London 93
   Local energy-efficient planning 95
   Energy-efficient planning in Milton Keynes 97
Denmark—integration of energy and spatial planning 101
Australia—an energy-efficient strategy for Melbourne 103
Conclusion 105

References 107

Index 117

Energy, spatial structure, and planning

Introduction
The dynamic interaction between energy systems and the spatial organisation of society has been a subject of considerable interest since the energy crises of the 1970s (for example, see Ashworth, 1974; Beaumont and Keys, 1982; Burchell and Listokin, 1982; Hall, 1979; Odell, 1977; Owens, 1984a; 1984b). At all levels of spatial resolution, from local to regional scale, energy systems influence spatial structure, and land-use patterns in part determine levels of energy consumption. An important implication of this relationship is that land-use policies may have significant consequences for future energy consumption. Land-use planners should be aware of these effects and could contribute, by appropriate policies, to energy conservation in the longer term.

Much has been said during the past decade about integration of energy considerations into the planning process; relatively little has been achieved, though there are some notable exceptions to this general rule in parts of the United States of America and in Scandinavia. There is certainly a need to take stock—to draw together what we know about the energy–land use relationship in theory, to identify clearly its policy implications, and to look at the successes, problems, and constraints of the limited experience of energy-integrated planning to date. These are the primary objectives of this book.

In the rest of this chapter, the links between energy and spatial structure, and the general case for including energy considerations in forward planning, are outlined. In chapter 2, the possible effects of energy constraints on the urban and regional trends of the late twentieth century are considered. These broad trends are influenced by forces largely outside the control of land-use planners, but are very likely to have an effect on energy requirements at the urban and regional scale. If energy efficiency is to become an objective of planning policy it is important to know whether other forces are working in its favour or against it. In chapters 3 and 4 the many possibilities for modifying urban form to reduce transport and space-heating requirements, respectively, are explored. Chapter 5 is a synthesis of the evidence on energy-efficient forms at different scales, leading to identification of important policy implications and constraints. In the final chapter practical experience in selected US cities, in Britain, and in Denmark, and an energy-conscious planning exercise in Australia, are evaluated. From this analysis interesting differences in perceptions and constraints in different political and institutional contexts are revealed, and important criteria for successful energy-integrated planning are identified.

In drawing together the theoretical and the practical in this way, the intention is not only to provide a much needed 'state of the art' summary, but to demonstrate the strong case for energy-conscious land-use planning and to show that, given the political will to overcome significant constraints, it *can* be achieved in practice.

Energy and spatial structure
The energy-land use relationship is represented schematically in figure 1.1. In reality, it is far from being so clearly defined; cause and effect are often difficult to distinguish and many aspects remain unquantified. Few planners in the United Kingdom, for example, would be able to find a documented 'energy budget' showing sources, flows, and end uses of energy in their county or local planning area. Statistics are usually available only at national level, even though energy budgets differ quite markedly on a smaller geographical scale and the potential for conservation and use of renewable sources is likely to show up more clearly in a 'fine-grained localised analysis' (Wilbanks, 1981). (There is surprisingly little work on this topic, but for interesting examples dealing with local and regional energy flows in Britain and elsewhere see Ball et al, 1981; Borg, 1981; Boyden et al, 1981; Jansson and Zucchetto, 1978; Zucchetto and Jansson, 1979).

The nature and availablity of energy sources clearly influence the spatial structure of society (link A, in figure 1.1). Historically, energy transitions—from the use of dispersed organic sources, through wind and water power, to large-scale exploitation of fossil fuels—can readily be linked with major shifts in settlement and transport patterns, culminating in the concentration of most of the population of developed countries into urban/industrial centres (see Lucasewiez, 1978; D Schumacher, 1985; E F Schumacher, 1976; Wrigley, 1962). It would be crudely deterministic to suggest that energy innovation *caused* these major social changes, but energy has undoubtedly been a crucial permitting factor. Cheap energy for transport and agriculture, for example, certainly removed two of the most important constraints on the growth of towns—the requirement that they should be sustained by an agricultural surplus and the transport 'bottleneck' arising from the need to be provisioned from ever more extensive surroundings (E F Schumacher, 1976).

If energy has been a permitting factor in the *process* of urbanisation, there can be little doubt that it has also exerted a profound influence on the location and direction of urban development during the twentieth century. Transport technologies and energy supply networks increased personal mobility, released industry from locational constraints, helped "to turn city dwellers into suburbanites" (Platt, 1983, page 27), and provided the fundamental permitting factors for urban growth to spread radially and at decreasing densities (Hall, 1978). [For fascinating case studies of the interaction between technical innovation, urban growth, and social change in Chicago, Denver, and Kansas City, see Platt (1983) and Rose (1983).] In short, there is ample evidence that energy supply, price, and distribution are among the important factors shaping urban and regional systems, even if the relationship is indirect and complex.

Society now faces the legacy of the most recent manifestation of this influence. Throughout a long period of falling real energy prices in the twentieth century the built environment has evolved, indeed has often been planned, in forms which reflect diminishing energy constraints in the transport system and the absence of concern for space-heating efficiency in predominantly low-density dispersed residential development. The result has been resource-intensive urban sprawl and increasing separation of activities in most Western countries.

The influence of energy on urban form is only one aspect of the relationship represented in figure 1.1. Once in place, land-use patterns and the built environment interact with the energy system in two important ways. First, they are among the determinants of the level and pattern of energy demand (link B). Spatial structure influences energy requirements for various activities, especially transport and space heating, which in the United Kingdom account for well over half of

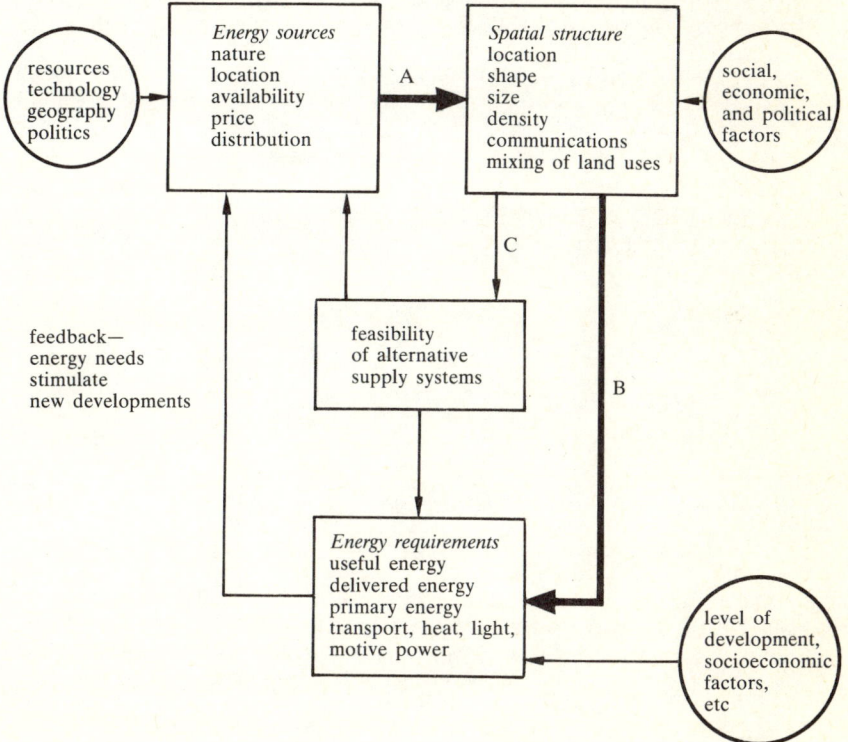

Figure 1.1. The relationship between the energy system and spatial structure.

delivered energy needs[1] (table 1.1). In low-density suburbs, for example, segregated from employment and services, and poorly served by public transport, people are necessarily dependent on a high level of personal mobility, and their travel patterns are inevitably 'energy intensive'. Second, spatial structure is an important determinant of the *feasibility* of future alternative systems for energy supply and distribution, such as combined heat and power generation (CHP) or the exploitation of ambient energy sources, which have particular requirements in terms of density, layout, and orientation (link C). Different aspects of spatial structure become important at different scales, from the regional, where the broad pattern of settlements is significant, to the local, where what matters is siting in relation to microclimate, layout, and orientation. Structural variables affecting energy requirements at different scales are shown in table 1.2.

Energy availability may change more rapidly than spatial structure can respond. In the period of steadily increasing energy availability and economic growth of the 1950s and 1960s, the built environment was able to respond gradually, and through growth, to the diminution of energy constraints. When the price of oil quadrupled virtually overnight in 1973, it was quite impossible for settlement structure, much of which

Table 1.1 Delivered energy requirements by end use in the United Kingdom (source: Leach et al, 1979).

End use	% delivered energy
Low-temperature heat (<80 °C)[a]	34.8
High-temperature heat (>80 °C)	25.0
Essential electricity[b]	8.1
Transport	21.2
Nonenergy uses[c]	11.0
Total	100.1

[a] Mainly for space and water heating.
[b] Lighting, machinery, electrochemical processes, etc.
[c] Chemical feedstocks, international bunkers, bitumen, lubricating oil, industrial spirits, etc.

[1] The distinction between primary, secondary, delivered, and useful energy is important. *Primary energy* is the calorific value of the fuel extracted from the resource base. It may be used directly (for example, coal for space heating) or converted to *secondary energy* (for example, electricity from coal), with conversion losses in the process. *Delivered energy* is measured at the point of delivery (for example, electricity or gas at the meter); it is a commonly used measure of 'energy demand'. *Useful energy* is that which is actually required to perform a function (for example, heat a room to comfortable temperature); it is less than delivered energy because of inefficiencies in equipment. In the United Kingdom, it is estimated that useful energy requirements are less than one third of total primary energy consumption (Ball et al, 1981).

had developed on the assumption of uninterrupted low-cost supplies of fuel, to adjust. The problem has, of course, been compounded by a much lower rate of growth and development, itself related to energy supply traumas, in the ensuing decade.

All of this—the close connection between energy and land use, inertia in the built environment, and the alarming prospect of inflexibility in the face of energy constraints—should add up to a strong case for integrating energy considerations into the land-use planning process. This was certainly the view in the 1970s, when all possible means of reducing energy consumption were on the political agenda, and planners were being urged to "re-examine all their precepts to see where they are based on the unlimited availability of natural finite resources" (Ashworth, 1974, page 777), to "consider the energy implications of their proposals with a view to adopting schemes of low energy intensitivity" (SCST, 1975, page 31), and to "devise and use methods for reorganising the ... activities of society into more efficient spatial relationships" (Brown, 1977, page 163).

But times have changed. Energy prices have fallen during the first half of the 1980s, and crisis reactions have become more considered, even dulled to complacency. The case for energy conservation is still compelling, but needs to be restated before looking at the special contribution which could be made by energy-conscious land-use planning.

Table 1.2. Significant structural variables at different scales.

Structural variable	Scale
Settlement pattern (for example, size and spacing of settlements, etc)	regional
Communication network between settlements	subregional
Size of settlement (area)	
Shape of settlement (circular, linear, etc)	
Communications network within settlement (radial, grid, etc)	individual settlement
Density	
Interspersion of land uses	
Degree of centralisation of facilities	
Layout (estates, etc)	neighbourhood
Orientation (of buildings or groups of buildings)	
Siting (in relation to microclimate)	building
Design	

The energy outlook in the 1980s

Projections of energy needs and resources have proliferated at all levels since the first oil crisis. They reveal more than anything else the uncertain and value-laden nature of energy forecasting. [For examples at the global scale see Eden et al, 1981; Foley and Nassim, 1981; Anderer et al, 1981 (and the important critique of this work by Wynne, 1984); IEA, 1985; Lovins, 1977; Odell and Rosing, 1983; D Schumacher, 1985; WEC, 1980. For the United Kingdom see Bending and Eden, 1984; Department of Energy, 1978; 1979; 1982; Leach et al, 1979.]

The most significant development in energy projections in the past decade has been the receding prospect of imminent physical scarcity of nonrenewable sources, especially oil. As demand forecasts have continually been revised downwards, the 'energy gap' in forward projections has narrowed or disappeared—the combined result of higher energy prices, recession, lower 'energy ratios'[(2)], and heavy investment in alternatives to oil (IEA, 1985).

But oil still supplies more than half of the world's energy and dominates medium-term prospects. There is consensus that in spite of a 'glut' of oil in the 1980s and strong downward pressure on prices, supply will continue to be 'tight'—for social and political reasons if not because of the size of the resource base itself (some 60% of world oil reserves are in politically sensitive areas in the Middle East). The substitution of other fuels (natural gas, coal, nuclear, and renewable sources) for oil is subject to technical, socioeconomic, and environmental constraints (see Eden et al, 1981; Foley and Nassim, 1981; IEA, 1985). Recently the International Energy Agency (a group of twenty-one Western industrialised nations) has warned against growing complacency, pointing out that without maintained and vigorous efforts to improve energy efficiency and develop alternative sources, "the IEA countries ... could once again become vulnerable to oil supply disruptions in the 1990's, similar to those of 1973–74 and 1979–80". The present 'easy' energy and oil markets increase this risk (IEA, 1985, page 12).

Not everyone agrees with the above analysis. Odell and Rosing (1983; 1984) present perhaps the best-known challenge to the conventional wisdom, arguing that recent developments in the world oil market have been misinterpreted, that forecasts have been too pessimistic, and that the current glut, far from being a temporary phenomenon, will last for several decades. This would provide a long 'breathing space' for a considered transition to alternative energy sources, eliminating the need to rush headlong into unfamiliar technologies. These projections are contrasted with some from a more typical analysis in figure 1.2. But energy efficiency is an important aspect of this analysis too, as a factor which has contributed to falling demand and which will be crucial in the

---

[(2)] The 'energy ratio' is strictly defined as the ratio between the growth rate of primary energy demand and the growth rate of gross domestic product (GDP). Sometimes the simple ratio between primary energy demand and GDP is used as an index of trends. The index of the ratio between total primary energy requirements and GDP for International Energy Agency (IEA) countries fell from 100 in 1973 to 81 in 1983.

maintenance of a more favourable supply-demand balance. Indeed, Odell (1975) was among the first to point to the importance of modifying energy-intensive settlement patterns as an integral part of energy policies in Western Europe.

Whether one takes an optimistic or a pessimistic view of oil resources, the case for conservation is not diminished by the abatement of immediate crisis. If anything, it has been strengthened as the negative sacrificial image associated with panic measures to reduce consumption whatever the cost—the 'freezing in the dark' option—has been replaced by a growing recognition of the considerable benefits of 'doing more with less'. Conservation is environmentally attractive [as energy supply facilities become increasingly contentious (Owens, 1985)], often a sound investment for the user, and may well (though this remains a controversial issue) represent a better investment for society as a whole than increasing energy supply (see SCE, 1981; Leach et al, 1979).

However strong the general case, particular energy conservation measures still need to be justified with reference to their benefits and costs (in the widest sense). This is always difficult, because of the many intangibles involved, and in the case of modifying land-use patterns, uncertainty is

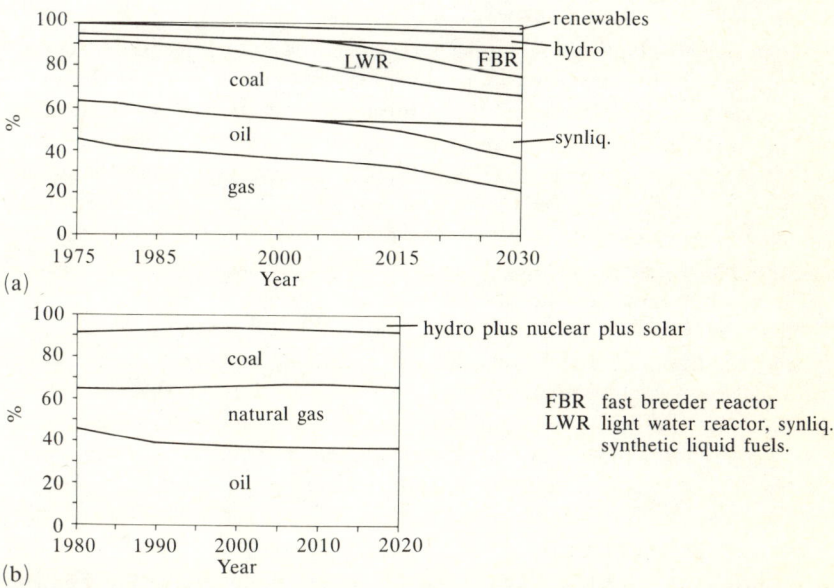

Figure 1.2. Comparison of energy projections. (a) The 'low scenario' considered by the Energy Systems Group at the International Institute for Applied Systems Analysis (Anderer et al, 1981). Total projected world primary energy demand in 2030 is about 16 000 million tons of oil equivalent, similar to that in Odell's scenario. The IIASA 'high scenario' has similar supply breakdown, but total projected demand is about 25 000 million tons. (b) Odell (1986) projects a much higher contribution from hydrocarbons and much lower from nuclear than does IIASA.

greatly increased because of the long time scale. Nevertheless, the argument in this book is that the benefits of energy-conscious land-use planning in terms of flexibility and, potentially, significant energy savings are considerable, whereas the costs are often small.

Energy and planning
It is highly undesirable for people to be committed to an energy-intensive life-style given the uncertainties surrounding future energy supply and prices. Furthermore, when current trends suggest that energy costs of projected urbanisation to the end of the century could amount to more than five times the 1973 world consumption of energy (Newcombe et al, 1978), it is clear that resource-conserving urban design will be crucial. Yet efficient and flexible land-use patterns are unlikely to evolve from energy-intensive ones without planning intervention, because of inertia, market imperfections, and the simple fact that land-use changes *are* subject to control in many countries. The case for including energy considerations in the land-use planning process is therefore both compelling and urgent.

Why, then, do many policymakers and planners, including most of those in Britain, continue to ignore this issue, or give it only limited attention? Is it simply a reflection of the fact that Western governments continue to place more emphasis on energy supply than on conservation? Is it because of the predominantly long-term nature of land-use planning policies and the slow turnover of infrastructure, presenting an inescapable physical constraint to modification in many areas? Or does energy efficiency conflict too much with social, economic, or environmental goals, or with powerful vested interests? All of these constraints will be shown to be important in particular situations, but the experience of those planning authorities which *have* tried to take energy into account shows that some of the obstacles are more apparent than real, and that many can be overcome, given sufficient political determination. The more intractable constraints apply to planning in general, not energy-conscious planning in particular, but few would argue that forward planning should be abandoned unless all constraints can be overcome.

Integration of energy considerations into the planning process is bound to be constrained by institutional inertia, as well as by the simple fact that the continually pressing demands of everyday activities militate against the careful consideration and incorporation of a new issue, about which little information has been readily available. This is especially the case in an era characterised by the 'retreat from government' (Cherry, 1982). In some British local authorities, for example, planning officers have found it difficult to persuade elected members that energy is important. But more commonly its neglect is due to the more immediate demands of other pressing problems, to the 'diffuse' nature of the energy issue, and to lack of knowledge and advice as to how integration could be achieved (Owens, 1986). Housing, employment, services, and the environment are predominant concerns, and the fact that the energy issue is so pervasive makes it difficult to see how it should be incorporated, because, as one planner argued, it is "not amenable to separate treatment under any single heading" (R Brown, personal communication, 29 November 1983). Another planner, responsible for a large and relatively deprived metropolitan area, said that his department "simply did not have the resources to deal with this complex issue on which virtually no government guidance or academic research was or is available" (P Wood, personal communication, 26 January 1984).

These comments suggest two things. One is that in Britain, at least, the fairly considerable literature on energy and land use is not reaching practicing planners, or is being rejected by them as too abstract; this in turn emphasises the need to demonstrate the practical significance of recent theoretical findings. The other is that the concept of energy as something linked to all aspects of land-use planning is not sufficiently appreciated. The pervasive nature of energy considerations should be a powerful argument for including them rather than an excuse for their continued neglect!

Given the links between energy and spatial structure, the need to use resources efficiently, and the fact that infrastructure provided now will influence energy demand for far longer than we can be certain about supply, it is irresponsible to ignore energy considerations in land-use planning. This does *not* mean that planners should embark upon a search for some energy efficient utopia forgetting that it is *people* who use energy and who live in 'spatial structures'. To urge that energy issues be considered in the planning process is not to imply that they should be given preeminence. Energy conservation is not an end in itself, but a means to allow society to prosper without the social, economic, and environmental costs incurred in the profligate use of a scarce resource. Land-use planners have a long-term role in contributing towards this end.

Spatial responses to energy constraints

Introduction
Historically, both the nature and the availability of energy sources have affected the spatial organisation of society at all scales. Most recently, the urbanisation, suburbanisation, and sprawl of the twentieth century may all be related in part to energy system changes, and particularly to the considerable growth in personal mobility made possible by the continued availability of cheap energy for motorised transport. It is interesting now to look towards the future and consider ways in which spatial structure might evolve in response to further changes in the energy system. Possible responses are discussed in this chapter, focusing mainly on the area about which most—though still much too little—is known, namely the influence of energy constraints on travel behaviour in the short, medium, and long term.

Reversal of trends?
It seems logical to assume that spatial structure will continue to respond to energy system changes, even if the direction of change is not in line with trends which have been apparent for most of the present century. Although there must inevitably be time lags in response because of the considerable inertia in the built environment, it is arguable that energy constraints will set certain forces in motion which will modify activity patterns and "open a veritable Pandora's box of spatial implications" (Wood and Lee, 1980, page 217). Simply stated, since the spatial organisation of society became notably more 'energy intensive' during an era of readily available fuel at decreasing real price, it seems intuitive that the trend at a time of energy constraint would be towards greater energy efficiency.

In the absence of much hard empirical evidence, views differ widely on the potential for spatial readjustments in response to energy constraints. At one extreme is the argument that
"The shape of the future metropolis may take a form determined *almost entirely* by energy availability: values and preferences may become submerged to necessity as our resource options disappear" (Van Til, 1979, page 321; author's emphasis added),
and at the other is the view of a recent US Presidential Commission that
"there are multiple options for accommodating higher energy costs... . Firms and households will likely be able to avoid profligate energy consumption in a variety of ways without resorting to relocation" (President's Commission for a National Agenda for the Eighties, 1980, page 32).
Which of these views is most likely to prevail? The simplest 'intuitive' scenario is one in which the most recent trends in spatial structure,

which have in effect been permitted by diminishing energy constraints, would be reversed if energy constraints increased in significance in future. For example, the trends towards decentralisation, lower densities, and the physical separation of activities, permitted by increasing mobility and also by the absence of any incentive to conserve heating fuel, might be halted and perhaps reversed. The underlying forces would then be leading towards increasing density, closer juxtaposition of homes, jobs, and services, and, in fact, to greater centralisation of activities once again. It has even been suggested that energy constraints, by stimulating such a reversal of trends, hold out some hope for the central city areas which have been so devastated by the exodus of population and employment. Newspaper and journal articles have talked graphically of 'imploding metropolis' and 'squeezing spread city' (Downs, 1974; Franklin, 1974).

The logic might be extended to speculate that if energy constraints became particularly severe, the larger urban areas may not be viable, since urbanisation itself is arguably a phenomenon related to energy availability. Janelle (1969), for example, saw urbanisation as part of the inexorable process of 'time-space convergence' as communications become easier, and E F Schumacher (1974) maintained that urban growth has been permitted by removal of two 'bottlenecks'—the need to service growing cities from ever wider areas, now met by cheaper transport, and the need for labour on the land, now increasingly replaced by nonhuman energy. Indeed, Schumacher and many environmentalists have argued that decentralisation and a return to or maintenance of small 'place bounded' communities (Hall, 1977) is the *only* way in which society will be able to live within an increasingly constrained resource budget. According to this view, some such change is therefore inevitable, and it would be preferable to bring it about in a deliberate and controlled way [for example, see the much publicised "Blueprint for survival" (*The Ecologist*, 1972) and E F Schumacher (1974; 1976)].

Intuition is often misleading, however, and the 'reversal of trends' scenario is almost certainly too simplistic for a number of reasons. One is that much will depend on the nature of any 'energy constraint'—a term too frequently used in this context without further qualification. Others include inertia in the built environment, the complexity of factors influencing spatial structure, and the multitude of options available for energy saving.

Response to energy constraints: some important issues
The nature of constraint
Energy constraints may take many different forms related to the price of fuel and to its physical availability and also to the source of energy and its distribution system. These different types of constraint may have very different implications for spatial structure. For example a smooth

gradual increase in the real price of oil, accompanied by gradual substitution of this fuel by other energy sources, would probably influence people and eventually spatial structure in a quite different way from a series of sharp increases in real price, each followed by a slow decline to a level rather higher than before (a pattern which has been characteristic in the United Kingdom during the past decade). Actual physical shortages of fuel (or the perceived threat of such shortages) may well have different implications from an increase in price, smooth or otherwise. This is not meant to imply that smooth changes in energy system parameters will necessarily result in similarly smooth adjustments to spatial structure. A recent application of catastrophe theory to the evolution of urban spatial structure suggests that gradual smooth changes in system parameters might in certain circumstances result in abrupt ('catastrophic') changes in the state of the system (Wilson, 1981). Simple models of retail facilities, for example, suggest not only that increasing energy constraints might lead to decentralisation in terms of size and location (a fairly predictable outcome, given the input assumptions), but that smooth changes in the deterrent effect of distance or consumers' scale economies could result in a sudden transition from one kind of retail system to another (Beaumont et al, 1981; Poston and Wilson, 1977; Wilson and Oulton, 1982). Quite how applicable this is to the real world remains to be demonstrated, which may be very difficult since real 'causal' factors rarely change smoothly over long periods of time. The point to be emphasised, however, is that individuals respond quite differently to different kinds of energy constraints, and that even gradual change may not result in an entirely predictable outcome (sounding a warning as far as policy is concerned). We still know very little about the nature of response to different kinds of energy constraint or about the possible eventual impact on spatial structure.

Not only the availability, but the nature of the energy sources which will eventually substitute for the nonrenewable hydrocarbon fuels of oil and natural gas, and the way in which the new sources are distributed, will impose their own constraints on spatial structure. Consider two rather extreme examples; an electrified society relying almost wholly on the centralised generation of electricity from nuclear power (the energy future implicit in much of the conventional wisdom about resources and policy) would be subject to very different spatial constraints from a society which had chosen to meet a significant proportion of its energy needs by exploiting renewable energy sources on the scale of the individual household or neighbourhood. Some of these points about energy constraints will be taken up in more detail later, but the intention here is simply to stress that the very nature of the 'energy constraint' may have important implications for spatial structure at all scales.

Inertia in the built environment

A second reason to question the simple 'reversal of trends' scenario is that the changes observed during most of the present century have taken place during periods of expansion and growth—in population, employment, service provision, and physical infrastructure. This infrastructure now exists, some of it still with a long expected lifetime, and by this simple fact it restricts what might be possible in terms of trend reversal, at least in the short to medium term. But it is also the case that there is much *more* infrastructure, serving a greater number of people and all their associated requirements, than there was, say, fifty years ago. The activities of industrialised society in the late twentieth century would not 'fit back in' to the compact centralised urban pattern from which they originally expanded.

Identifying 'causes'

A third and probably most important factor which militates against reversal of trends in the face of energy constraints is that, for the most part, changes in spatial structure which have resulted in energy-intensive land-use patterns have been permitted but not *caused* by the availability of plentiful energy at a low price; the energy system has been more analogous to a catalyst than a prime ingredient in the process. For example, a complex of social, economic, political, and technological factors—itself much debated—has been behind the outward migration of residents and firms from the inner cities. It has been reinforced by negative feedback in the inner areas, where loss of the more prosperous activities has led to loss of revenue, ever decreasing quality of life and opportunity for those who remain, further out-migration, and further decline. In contrast, there is positive feedback in the suburbs (and, increasingly, in small towns away from the metropolitan area), whose desirability is reinforced by the continued decline of the central areas. This is not the place to delve into the debate about the causes of inner-city decline but it must be recognised that the forces underlying it may prove irresistable even in the face of strong energy constraints unless there are other social changes. And, even if energy system changes do lead to spatial readjustment, this may not constitute a simple reversal of trends and the return of people and jobs to the inner-urban areas. Similar arguments apply to other spatial changes which have been influenced, but not caused, by energy-related factors in the past.

Nonspatial options for energy saving

The final and related point is that energy is only one of many commodities purchased by individuals and institutions. A rise in the real price of fuel may or may not result in consumption of less fuel. If the demand is inelastic—for example, if the energy consumed in the journey to work is seen to be indispensible—consumption of some other commodity may be given up instead. If the demand *is* elastic, it may be reduced in ways which need not have any spatial implication, such as a switch to a more efficient car for the journey to work. Thus it is possible that adjustments to energy constraints will be made in ways which will not be manifest in any forces to change spatial structure, and certainly a return to more compact urban patterns is not inevitable. It should be noted, however, that there is some evidence from surveys conducted after the 1974 oil crisis in the USA to suggest that fear of periodic *nonavailability* of energy for the journey to work might encourage relocation to a greater extent than would fuel price increases (Corsi and Harvey, 1977).

The situation is a complex one in which the extent to which 'the market' will cause spatial structure to be modified in response to energy constraints is not at all clear. A critical factor will be the relative elasticities of energy demand for those purposes which can be related to spatial structure and for those which cannot. The complexity of the situation is compounded by a dearth of empirical evidence on the ways in which people actually respond to energy constraints, either financial or physical. What we are left with then is a sense that future changes in the energy system are likely to influence spatial structure, but considerable uncertainty about how, and to what extent. Neither theoretical nor empirical approaches to this question have proved entirely satisfactory.

Energy constraints, behaviour, and spatial structures
It is possible, of course, to incorporate energy coefficients into standard models of urban structure and then to explore the effects on structure of changes in the energy system parameters: indeed, energy costs are implicitly included (as fuel costs) in the measure of impedance employed in most urban models. The problem is that modellers have tended to *assume* that aspects of individual behaviour which might ultimately be reflected in spatial structure are elastic with respect to energy constraints. Given this assumption, the results of the exercise become largely predictable. But there are many ways in which people might respond to increasing energy prices, or fuel scarcity, which do not have obvious spatial implications. We need a much better understanding of actual behaviour, but attempts to unravel its complexity have, so far, shown a "frightening lack of uniformity" (Dix and Goodwin, 1982, page 189). These attempts—and some tentative conclusions—will now be considered in more detail, before going on to discuss some of the more formal urban-modelling exercises and their results and limitations. The discussion must inevitably focus on the transport sector, since it is here that the most significant insights into responses and their structural implications have been gained, but some consideration is also given below to possible effects on the housing market.

Response to fuel price increases in transport
Evidence about individual response to energy constraints appears at first sight to be ambiguous and contradictory. The reason lies largely in the absence of smooth trends in real fuel prices since the early 1970s, against which adjustment behaviour could be measured. Changes in price in one direction (up) have typically been sudden and large, whereas changes in the other (down) have been gradual, resulting from inflation (Dix and Goodwin, 1982). In these circumstances it has been very difficult, for example, to measure demand elasticities for transport fuels. Nevertheless, attempts have been made to measure elasticities, and surveys have been conducted which provide some indication of people's likely response to various types of energy constraint. What follows is based on this rather limited information.

Envisage two alternative scenarios. In the first, motorists respond to increasing petrol prices by decreasing car use. In the short term, they cut down on 'unnecessary' journeys such as social and recreational trips, but in the longer term they attempt to rationalise their trip-making patterns, for example, by changing employment or by moving closer to work or to a service centre. In the second scenario motorists maintain levels of car use but contain expenditure in the short term by driving more carefully or by reducing expenditure on other goods. In the longer term they buy smaller and more fuel-efficient cars, thus maintaining mobility but using less energy. Clearly, the first scenario has significant implications for car use, petrol consumption, and spatial structure, whereas the second will influence petrol consumption (at least in the longer term) but not car *use* or the spatial arrangement of activities (except in the wider sense because it will influence demand for energy, cars, and other goods). There are of course many other possible combinations of events and responses, of which these two scenarios represent different extremes, but for anyone interested in possible spatial adjustments to energy constraints, it is clearly important to know whether reality is likely to be closer to the first scenario or to the second.

The available evidence comes mainly from studies whose original purpose was the prediction of levels of ownership and use of different classes of vehicle, to provide an important (and controversial) input into the transport policy process. From such studies it is possible to glean some implications for future spatial adjustments to energy constraints, though it must be said that the findings, on the whole, are too contradictory and ambiguous to give any really clear indication of the likely future evolution of land-use and activity patterns.

The "frightening lack of uniformity" in studies of response to petrol price changes arises partly from the diversity of analytical frameworks (including, for example, the 'economic' and the 'psychological') and partly because of the different time scales over which adjustments have been considered. Dix and Goodwin (1981; 1982), having reviewed most of the relevant work, suggest that apparently incompatible views could be reconciled by considering different time scales and by distinguishing between the effects of increasing prices on petrol sales and the effects on traffic levels (the latter, of course, having more direct relevance in the longer term for spatial structure).

Most evidence suggests that petrol price increases are unlikely to have a very marked effect in the short term. Marginal adjustments of trip patterns, especially social and leisure trips [for which elasticities have, not surprisingly, been shown to be greater than those for more essential journeys (for example, by Lewis, 1977)], are associated with only small changes in petrol consumption. Short-term elasticities of around -0.25 are relatively well supported by available information (Department of Energy, 1977), and most lagged time-series analyses show higher long-term than short-term demand elasticities (especially when demand is measured by petrol sales). Evidence that people adopt coping strategies in the short term is also supported by 'backtrack' interview surveys of adaptation to cost changes, many of which were conducted in the aftermath of the fuel crisis of the 1970s (for example, Carpenter and Dix, 1980; Corsi and Harvey, 1977).

In the medium term, as shown in figure 2.1, 'traffic' and 'petrol' elasticities seem to diverge, the former becoming even smaller as people try to revert to their former trip patterns, and the latter increasing as vehicles are changed to save petrol without loss of mobility (Dix and Goodwin, 1981). Clear evidence for this kind of response is provided by an analysis of household expenditure on transport in the United Kingdom in the early 1970s (Mogridge, 1977). It suggests that the short- to medium-term response would *not* be one of spatial readjustment. It may not even be rational, since perceptions of travel costs and trends in

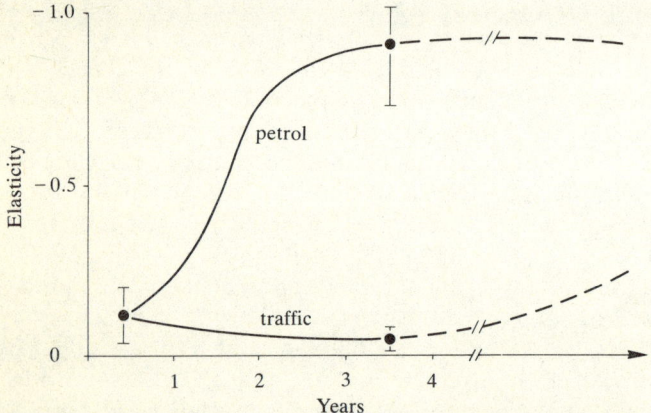

Figure 2.1. Dix and Goodwin's 'reconciliation hypothesis' (source: Dix and Goodwin, 1982).

petrol prices are notoriously inaccurate, hardly better in many surveys than a 'distribution of guesses' (Metcalf, 1978).

For longer term spatial readjustments to occur, 'traffic' elasticities would have to rise again as decisions on location and life-style were influenced by energy constraints. Dix and Goodwin (1981) argue that this is indeed a possibility, since in the *long run* people adjust rationally to costs, even if they do not do so in the short term, and will strategically adjust their travel habits around the time of 'life shocks', such as important transitions between life-cycle stages. Although there is some support for this hypothesis [from French work on elasticities, for example (Marche, 1980—discussed by Bates and Roberts, 1982)], it is really no more than informed speculation about human behaviour. However plausible the view that energy constraints must ultimately effect spatial adjustments, no one has yet provided clear empirical evidence that this is likely to be the case. Indeed, such evidence may be elusive, for as a Department of Energy Working Group has pointed out,

"The availability of more data in future may enable us to improve our estimates, but elasticities ... are liable to change over time and we are unlikely ever to arrive at a single final answer" (Department of Energy, 1977, page 10).

Physical shortfall

The discussion so far has centred on the effect of fuel price changes. For obvious reasons, there is even less opportunity to study the impact of a shortfall in the physical availability of petrol. A rare chance to do this has arisen, however, in Tasmania, where the destruction of the only cross-river bridge in Hobart left some 30% of the urban population (40 000 people) effectively isolated from centres of employment and services. Reconstruction took three years, during which time it was possible to study the various ways in which residents responded to the situation. In general, results are similar to those found elsewhere in relation to price increases. Operational adjustments which were easily implemented, reversible, short-term, and noncommitting tended to be invoked before structural adjustments involving more substantial and more permanent action, such as locational change (Wood and Lee, 1980; see table 2.1).

Bearing in mind the relatively short time scale involved, and the points already made about adjustment at the time of 'life shocks', it is perhaps not surprising that structural changes in Hobart were postponed; obviously the *expectation* that the bridge would be restored as soon as possible would have been an additional factor encouraging nondrastic coping strategies. Expectation of repeated physical shortfalls might, on the other hand, act as a strong additional factor to encourage relocation at the time of life-cycle changes. Although they conclude that in the short term significant changes to spatial structure seem unlikely, the

Hobart investigators propose that in the longer run "a major consequence of restrictions on mobility is likely to be a contraction of activity spaces and an increase in the degree of self-containment of sub-sections of urban areas" (Wood and Lee, 1980, page 220).

Table 2.1. Some adjustments to mobility constraints in Hobart (source: Wood and Lee, 1980).

Type of activity	Operational[a]	Structural
Travel to work	adopt car pooling adoption of flexitime to minimise peak congestion	permanent relocation of job permanent relocation of residence (purchase of second residence) (pied-à-terre) withdrawal from workforce
Shopping	reduce frequency—more multipurpose trips increase phone orders alter shopping role within household	change to more proximate shopping location
Travel for medical treatment		change to more proximate medical practitioner
Social interaction	accept lower frequency of contact increase phone contact	redirect social ties on local areas
Leisure/recreational outings	accept lower frequency of outings	redirect outings to local facilities increase home-based leisure activities

[a] Changes in travel mode and route were operational adjustments evident for all kinds of activity.

Summary

In summary, the consensus seems to be that, at least in the short to medium term, people will respond to energy constraints in ways which will result in less energy consumption whilst having little significant impact on spatial structure. Empirical evidence on the nature of longer term adjustments is virtually nonexistent, but the possibility of relocation to reduce travel requirements has been seriously considered by a number of authors and should certainly not be ruled out. It is difficult, at this stage, to come to any more definite and satisfying conclusion.

Energy constraints and urban models

It should now be apparent that we cannot simply *assume* that people will travel less in the face of energy constraints. But it remains a plausible hypothesis that individuals might adjust their travel habits in the longer term in ways which will ultimately have an impact on spatial structure. If they do, how might spatial structure be modified at different scales? How would responses to energy constraints interact with other trends in the evolution of new patterns of spatial organisation?

Once the assumption of 'elasticity' in travel patterns has been made we might expect urban models to be of some help in exploring these important questions. In fact, although various models have quite frequently been used in attempts to do just this, they have so many obvious limitations that the results can be interpreted only with considerable caution.

Many of the criticisms of urban models are well known, particularly those relating to their gross oversimplification of reality and the common assumption that what is essentially description can be used for prediction without any intervening understanding of the social processes involved. Such criticism would of course apply to models which 'predict' spatial adjustments to energy constraints by adjusting those parameters which are related to energy costs. It is not surprising, for example, that increasing the importance of transport costs in the 'trip decision' in a Lowry model should result in closer association between different activities and a reduction in the total amount of travel (Beaumont and Keys, 1982); or that a utility analysis of residential location should suggest that higher energy prices are likely to result in residential areas closer to employment centres (Romanos, 1978). Similarly, at the intraurban level, it might be anticipated that a simple geometric model of an urban area combined with a standard spatial interaction shopping model would predict that energy constraints might lead to decentralisation of retail facilities in terms of size and location (Beaumont et al, 1981). In a sense, these models simply formalise the 'intuition' that if physical separation of facilities increased in the absence of energy constraints, it will decrease again if energy becomes more scarce and expensive. But as Mogridge (1984, page 592) points out in an outspoken critique of such work, "it is quite useless simply to change the deterrence function parameter from a low value to a high value and show that the city alters its form. What we need to know ... is whether the effect of a change in transport cost can be accurately modelled". This reinforces the need for a better understanding of behavioural responses to energy constraints.

Another problem is that for many of the modelling exercises, somewhat outdated assumptions are made about *existing* spatial structure or about important current trends. For utility analyses, such as those used by Romanos (1978) and Dendrinos (1979), for example, it is assumed that the residential-location decision by 'households' is the result of a trade-off between accessibility and space or amenity, the former decreasing and the latter increasing with distance from 'the centre', where employment and service facilities are assumed to be concentrated. If the centre is assumed to be the traditional central business district and households are assumed to be uncomplicated families with one breadwinner, then naturally enough an increase in energy costs in these models will result in greater centralisation, especially if, as in Romanos's model, residential space-heating requirements are assumed to increase with distance from the centre (because of their relation to dwelling densities).

As soon as other factors are even rather crudely introduced, for example, time costs equivalent to income foregone during commuting, the outcome becomes less predictable and the city does not *inevitably* become more compact in the model (Dendrinos, 1979). Romanos did at least build the suburbanisation of employment centres into his analysis, with the result that instead of a single, compact, centralised urban area, his model predicts a pattern of 'decentralised concentration' as the emerging equilibrium in a situation of energy constraints. Again, this is not a surprising outcome given the assumptions of the model, but it is interesting to see the decentralisation and nucleation of activities being postulated as the result of 'time-space divergence' (Wood and Lee, 1980). Janelle (1969) in his original work on time-space connectivity also predicted the emergence of this spatial pattern, but only as an intermediate stage in the inexorable process of time-space *convergence*, leading towards an expanded area of centralisation and specialisation.

Nowhere has the potential influence of energy constraints on the more recently observed phenomenon of 'counterurbanisation' been explored, yet this is perhaps the most important trend in the location of human activity in the USA and the United Kingdom, and is observable in many other advanced countries (Berry, 1976; Fielding, 1982; Robert and Randolph, 1983). Counterurbanisation involves the net flow of people and jobs out of the major conurbations not into the suburbs but into small- and medium-sized towns of peripheral and rural regions. In so far as this implies the growth of settlements well away from and largely independent of the conurbations, it may well represent a more energy-efficient pattern than does suburbanisation (which remains a significant process) (figure 2.2). Energy constraints may reinforce this trend, as well as making suburban centres more autonomous; and, therefore, far from reversing the exodus from the metropolitan areas, energy constraints might actually encourage it. Could we even detect here the beginnings of the return to 'smaller place-bounded communities' which has already been proposed as an appropriate spatial policy for an energy-constrained future?

Models tend to indicate that energy constraints would lead to a reduction in the physical separation of activities, brought about by higher densities or decentralisation of facilities or both; but it must be remembered that these conclusions are largely built into the models by the assumptions on which they are based. One thing which they do help to clarify, perhaps, is that *if* some of the assumptions about individual responses are correct, the foci of emerging spatial structures are likely to be existing centres, and that these are perhaps more likely to be in increasingly autonomous suburbs or in areas well away from the conurbations than in the old traditional city centres themselves.

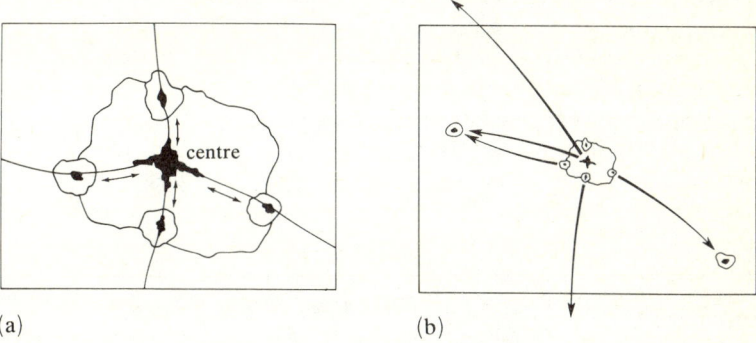

(a)   (b)

Figure 2.2. Suburbanisation and counterurbanisation. (a) Suburbanisation: some autonomy in suburban centres, but links maintained with metropolitan centre—energy intensive. (b) Counterurbanisation: loss of people and jobs from the whole of the metropolitan area, growth of autonomous smaller towns (the 'clean break')—energy efficient?

Residential energy consumption and spatial structure

Changes in energy availability are also likely to affect residential energy consumption. Space heating (and cooling, in some climates) accounts for the largest single part of domestic energy use and is the aspect which is more directly related to spatial structure. In relation to market responses and possible changes in spatial structure the situation for space heating is more complex than for travel, since fuel substitution is possible to a much greater degree. Households can respond to price increases (or shortfall in supply) by switching to a different fuel, rather than by decreasing the total amount of energy consumed. (The effect of changes in the price of one fuel on the quantity consumed of another fuel is expressed formally in terms of cross-elasticities of demand.)

The long-run price elasticity of residential energy demand (averaged over nine countries) has been estimated to be -1.1, with own price elasticities in the region of -1.0 for individual fuels [Pindyck (1979); a range of estimates is given by the Department of Energy (1977)]. What is of interest in the present context is whether energy constraints in the residential sector might induce a response in individual households which will be reflected in the housing market and, ultimately, in spatial structure. Whether this occurs will depend on the particular ways in which people adjust to increases in the price of fuel for space heating. They may, for example, substitute fuels, tolerate lower temperatures, or eliminate waste, none of which will necessarily have any influence on spatial patterns. But it is also conceivable that the energy advantages of more efficient built forms will be recognised. At the local scale, factors such as siting in relation to microclimate, the physical arrangement of buildings, and built form can make a very significant difference to space-heating requirements. An intermediate flat, for example, could be three times more efficient in this respect than a detached house of similar size; and a house which is carefully orientated to take full advantage of passive solar energy could have half the heating bills of one whose siting ignores this potential.

The interesting question is whether recognition of energy-related characteristics at the 'microscale' might result in increased demand for energy-efficient buildings, inducing migration to areas where they predominate and influencing the built form and density of new construction. This could in theory encourage a movement inwards towards city centres (where terraces and flats are most likely to be located); or, as prices rise, people might turn increasingly to semiautonomous energy supply systems, such as solar space and water heating, as these become more economically attractive. This too might encourage migration, but in a different direction towards lower density areas where such systems may be most viable (Owen Carroll and Udell, 1982), and would influence the built form, layout, and density of new construction. Indeed, this influence may already be observed in some areas, such as Southern California, where builders are responding to growing demands for solar 'access', which in turn influences the spatial structure of new developments.

Generally speaking, however, there is as yet little evidence that energy considerations are influencing migration decisions, or that they are having very much effect on spatial structure in new developments (except, as in the Californian example, at the smallest level of spatial resolution). There has been virtually no response of this kind in the United Kingdom. Although it seems that home purchasers are giving increasing consideration to energy efficiency (for example, see Kaiser et al, 1982), and may selectively purchase houses with cheaper heating systems (for example, see Halvorsen and Pollakowski, 1979), they have not yet expressed a preference for higher (or lower) density living for reasons of energy conservation. And this is hardly surprising, given the way in which the housing market operates. Terraced houses and flats may be more energy efficient than detached dwellings, but they are also usually smaller, cheaper, and serve a different sector of the market. The perceived disadvantages of higher density living are clearly reflected in most estate agents' windows. Unless some changes occur, or can be brought about, in people's preferences, it seems unlikely that the energy advantages of being 'joined on' will outweigh the disadvantages such as noise, and lack of privacy. Presumably, houses will be built to reflect these preferences, and people's ability to pay for them. This may seem obvious, but it is surprising in how many theoretical and empirical studies, energy-efficient built forms are specified without reference to the fact that historically and for easily identifiable reasons, such forms have not been preferred and, in general, 'moving up through the housing market' has meant moving to less energy-efficient built forms. To show that significant energy savings can be achieved by what, for many people, would be a diminution of the quality of life is the kind of analysis which gives conservation a bad name. If changes in life-style towards higher density living, with less space, are considered *desirable* and regarded in a positive light, that is another matter; and it may of course be argued that greater efforts should be made to overcome the known disadvantages of higher density so that its energy advantages may be realised. Siting and orientation are rather different considerations, since major energy savings without commensurate disadvantages are demonstrable. But these can be exploited only in new construction, and many constraints—including the slow rate of turnover of the building stock and the innate conservatism both of construction companies and of their customers—will limit any short-term influence on spatial structure.

In conclusion, it must be said that since there are so many ways in which residential energy consumption can be reduced, the fact that energy considerations may be increasingly important in house purchasing decisions is by no means a sufficient condition to effect significant longer term changes in spatial structure. However, as with the adjustments which might be brought about because of changes in travel behaviour, modifications arising from residential energy considerations should not be discounted in the longer term.

Concluding comments
To try to draw these various threads together to reach a conclusion is difficult and rather unrewarding, for we do not know enough about the energy future or about the way in which people might respond to price rises or shortages to be able to predict the influence on spatial structure with any confidence. It cannot be assumed that people will be prepared to relocate to reduce energy consumed in transport or in the home, but if they do, evidence suggests that the longer term effect is likely to be closer integration of different activities. Superimposed on current urban trends, this may lead to greater autonomy in urban subcentres and in smaller freestanding towns—'decentralised concentration'. Further research may help to confirm or contradict these rather speculative ideas, but even a much clearer understanding of the present situation (in relation to price response, for example) may only slightly reduce the hazards of prediction. In the end, however, it is not accurate prediction which will matter as much as flexible normative planning, based on an intelligent assessment of the most likely direction of certain trends. Having looked at 'what might happen', therefore, it is now time to turn to the question of how we might plan, spatially, for an energy uncertain future. This is the question to which the following chapters are addressed.

Reducing transport energy requirements

Introduction

The two most important ways in which spatial structure and energy systems are related are through travel and transport requirements and through energy use in buildings, mainly for space heating. In so far as land-use planners influence transport policy and the density, form, and siting of new development, they are presented with an opportunity gradually to improve the energy efficiency of the built environment. This opportunity raises two fundamental and important questions.
(1) What should planners aim to achieve?
(2) How effective are any energy-related planning policies likely to be?

To answer the first question, we must identify energy-efficient structures at different levels of spatial resolution. The second question is addressed in the process, for it rapidly becomes apparent that although potentially efficient forms can be identified, the achievement of actual energy savings will depend on many factors which planners do not necessarily control.

Approaches to defining energy-efficient structure

Energy-efficient spatial structures will be those in which *useful* energy requirements are minimised (subject of course to other constraints) and/or those where energy needs can be met in ways which minimise the ratios along the chain of primary, secondary, delivered, and useful energy[3]. In practice, energy efficiency will be achieved by some combination of these factors. We should aim, for example, to identify spatial structures which reduce the *need* to travel while at the same time providing a configuration of land uses compatible with energy-efficient modes of transport. Similarly, it should be possible to combine built forms which are intrinsically energy efficient with land-use patterns that are compatible with, for example, district heating systems.

Current knowledge of what constitutes energy-efficient spatial structure comes from a wide variety of sources, not all of them directly concerned with energy use or conservation. These can be divided into five broad categories, though there is considerable overlap between them.

First, there are studies, already described, of the likely response of spatial structure to energy constraints. On the assumption that equilibrium would tend to be restored by a trend towards greater energy efficiency, the emerging land-use patterns in these studies provide some insight into energy-efficient characteristics. Assumptions, however, have tended to be heroic and conclusions understandably tentative.

Second, there is empirical comparison of energy consumption in different geographical areas. To the extent that differences in energy-consumption patterns can be correlated with structural variables, some

---

[3] See footnote 1, chapter 1.

conclusions may be drawn about the energy implications of different spatial structures. This apparently simple approach to investigating the energy-urban form relationship is, however, beset by numerous difficulties. From the outset, there are problems of defining and quantifying the various dimensions of urban form in existing areas, which are after all only the current stage of a long process of evolution: they may not reflect any specific urban form and certainly cannot represent all possible urban forms (Stone, 1973). Then, considerable variation in energy demand patterns due to climatic, socioeconomic, and many other factors tends to obscure any variation which might be attributed to spatial structure. Finally, great caution must be exercised in claiming causality, even when a high correlation between energy consumption and structural variables is observed. Empirical work has mainly been concerned with transport energy requirements, although some studies have also included residential energy consumption. Results have tended to be inconclusive with significant relationships not always emerging clearly and with many contradictions between different studies.

A third approach, and the one which has most commonly been employed, is to estimate the energy requirements associated with alternative hypothetical spatial patterns and to evaluate the alternatives in terms of energy efficiency. This approach has the advantage over empirical work that alternatives can be investigated in 'laboratory' conditions, with variables other than those of direct interest being held constant. However, it invariably involves urban modelling and is thus open to all the criticisms which can be levelled at this technique, especially, in the present context, the need to oversimplify the real world often to such a degree that it is difficult to see how the results could have any genuine policy application.

Another problem relates to the fact that alternative spatial configurations for evaluation are selected with varying degrees of subjectivity. Although the majority of theoretical possibilities will be ruled out by nonenergy considerations in the real world, it is arguably desirable in exploratory studies to include as wide a range of alternatives as possible. Subjective selection permits examination of a small number of options in detail, but carries the obvious risk that some potentially energy-efficient structures will be overlooked. At the other extreme, if the objective is to evaluate all possible hypothetical structures, the models used must be extremely simple.

Theoretical studies have ranged from the highly abstract (for example, Hemmens, 1967; Schneider and Beck, 1973) to those which consider alternative development patterns for existing geographical areas (for example, Clark, 1974; Roberts, 1975), though even the latter have been highly simplified and abstracted. Examples of some of the different

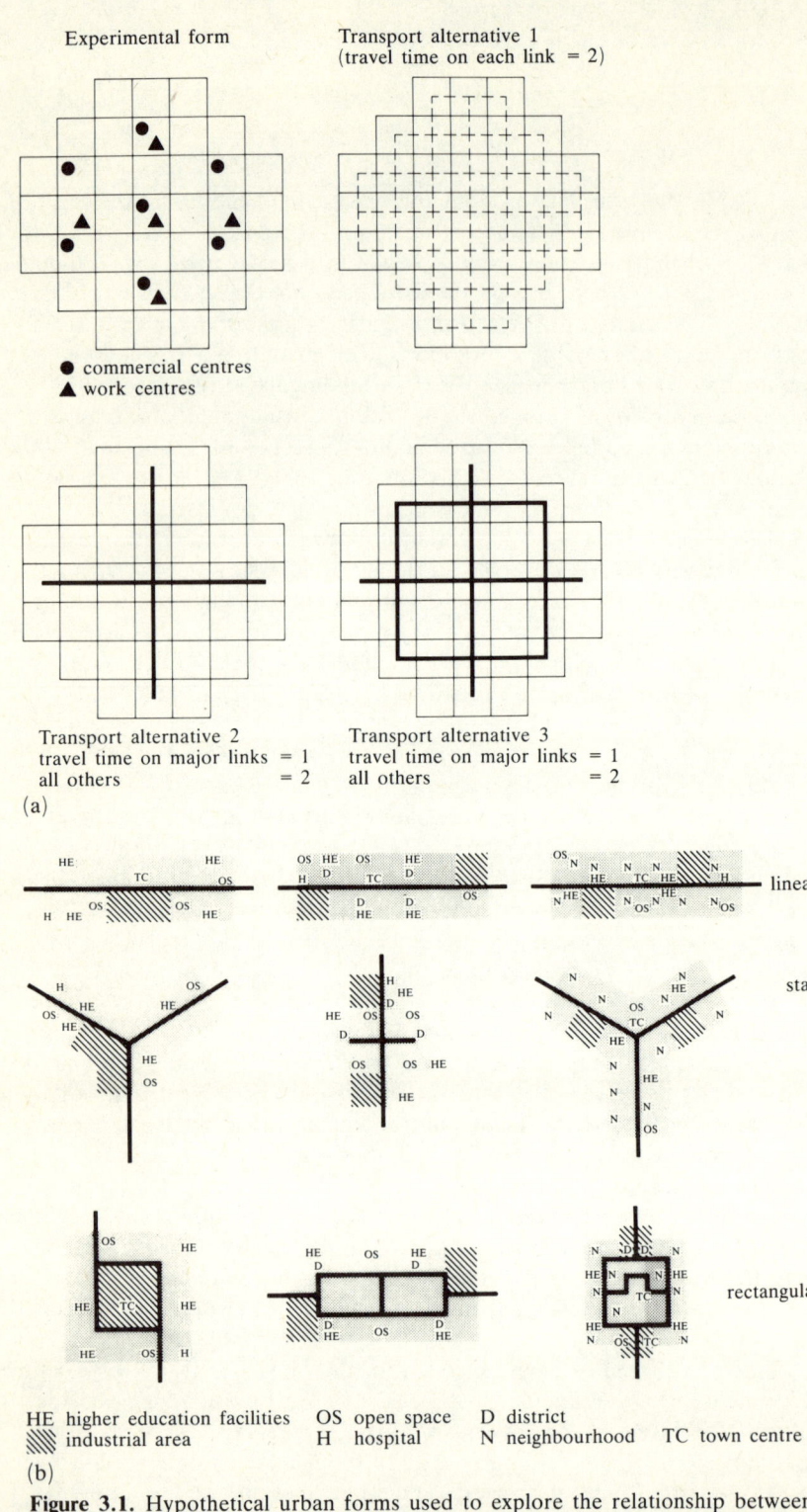

Figure 3.1. Hypothetical urban forms used to explore the relationship between spatial structure and travel/transport energy requirements by (a) Hemmens (1967), (b) Stone (1973), (c) Edwards and Schofer (1975).

forms investigated are shown in figure 3.1. Again, the emphasis has been on the transport sector, with investigators most typically using gravity type models to 'predict' the travel and energy demands associated with alternative arrangements of land uses.

A fourth and more normative approach is to identify the spatial *requirements* of energy-saving techniques, such as combined heat and power generation, use of renewable energy sources, or promotion of public or nonmotorised transport. These requirements may be derived from technical considerations, models, and working experience. The emphasis here tends to be on reducing energy requirements for heating in buildings, but there are also some important transport implications.

Last, there is much to be learnt from combination and extension of the above methodologies into normative planning and evaluation of specific ideas.

The remainder of this chapter focuses on travel needs and transport energy requirements, and on ways in which these might be reduced by appropriate land-use patterns. Transport accounts for some 16% of total primary fuel consumption in the United Kingdom and, more

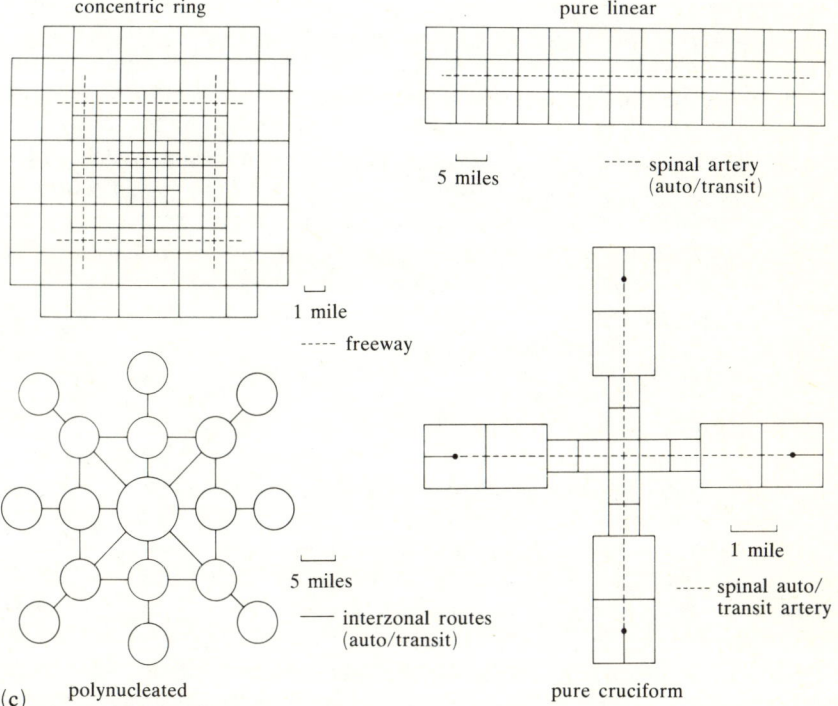

Figure 3.1. (continued)

significantly, is almost entirely (99%) dependent upon oil. Energy use in buildings is dealt with in chapter 4. Before proceeding, however, two points must be emphasised. The first relates to the significance of scale: it is always important to bear this in mind since different structural factors influence energy requirements at different scales. The second point, perhaps even more significant but often overlooked, is that although it is necessary and interesting to identify the characteristics of an energy-efficient environment starting on a 'clean state', it is also of crucial importance to identify the kinds of incremental modifications to existing spatial structure which would make it more energy efficient. In the developed world particularly, planners are more likely to be interested in energy-efficient directions for modest growth and change than in ideal structures and policies for new settlement planning on a large scale.

Reducing transport energy requirements
There are two basic ways in which energy needs for transport can be reduced—people can travel less and they can travel by more energy-efficient means. The former implies arranging land uses so that there is less *need* to travel and the latter implies aiming for land-use patterns which are most conducive to public and nonmotorised transport.

Most of what is known about spatial structures and transport energy requirements applies to land-use patterns at the regional or subregional and the intraurban scales, especially the arrangement of land uses within individual urban areas. Relevant variables are the size and shape of the communications network, density of development, and interspersion of different activities. In many empirical and theoretical studies, the effects of these variables on travel or transport needs have been examined, and some work—mainly at the intraurban scale—has focused on defining spatial characteristics which are most compatible with the operation of energy-efficient transport systems.

The regional and subregional scales
The questions of how large individual settlements should be and whether there is an ideal shape for an urban area have occupied planners for rather longer than their concern with energy efficiency, but clearly they have important energy implications. However, at this level of analysis— the regional and subregional scales—evidence about 'efficient' spatial structure remains sparse and no really satisfactory conclusions can be offered. Perhaps this is because there is always a faint air of unrealism surrounding any attempt to identify 'ideal' patterns of any kind at this scale. The broader the scale, the less real influence land-use planners are likely to be able to exert on development patterns. Cherry (1982)

suggests, for example, that

"in ... the market economy countries of the West, public sector land policies have remarkably little effect at macro scale on the shaping of metropolitan form and regional distribution patterns" (page 148).

In the United Kingdom there is no land-use planning at any scale greater than that of a county or a metropolitan county (region in Scotland). It is hardly surprising then that ideal spatial patterns at the regional scale remain less well-defined than those at a more manageable level. Nevertheless, the questions of size and to some extent shape of urban areas in relation to travel requirements retain some significance, and it is worth examining them in more detail before moving on to the more immediately relevant factors at the intraurban scale.

Urban size

The question of whether travel needs increase or decrease with urban size remains unresolved. It is not possible to identify an 'ideal' urban size from the point of view of transport energy requirements, or even to say with any conviction whether their relationship with size is negative or positive. The uncertainty and contradictions in empirical evidence arise largely from the multitude of factors which intervene in this relationship, making it very complex.

In the United Kingdom, per capita energy use for all transport (and for private transport alone) declines with size of urban area from small towns to provincial conurbations (Maltby et al, 1978). But Greater London is an exception to the trend, having the highest per capita energy use of all the urban areas. A similar pattern emerges for per capita distance travelled. These variations are illustrated in figure 3.2, which

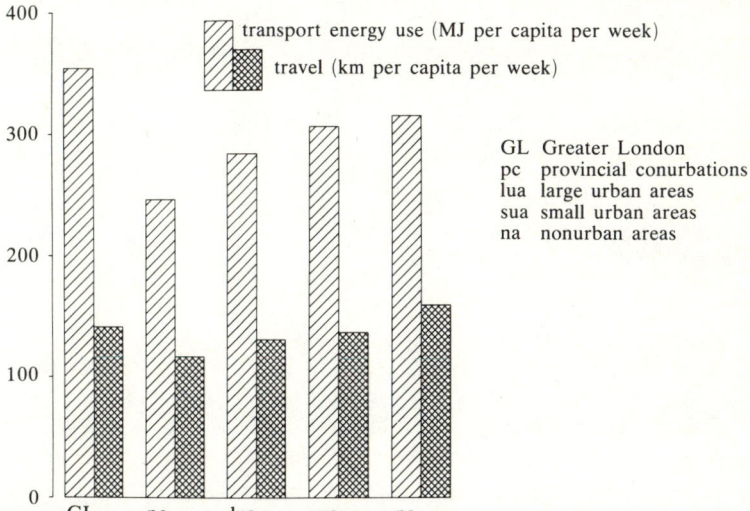

Figure 3.2. Travel and energy use for transport by different types of area, 1972 (source: from data in Maltby et al, 1978).

also shows the relatively high travel and transport energy requirements in nonurban areas (settlements with less than 3000 inhabitants or a dispersed population). Interestingly, there seems to be little variation in the total *number* of journeys made per head of population by residents of different areas in the United Kingdom, suggesting that travel mode and journey length account for differences in energy consumption.

Theoretical work on the other hand, also based on towns in the United Kingdom, provides evidence of a *decline* in transport energy efficiency with urban size, but the results need qualification. In a comprehensive theoretical analysis of the effect of urban form on the financial costs of construction and operation of settlements, the National Institute for Economic and Social Research found that journey-to-work costs (estimated using a model based on a gravity formula) increased

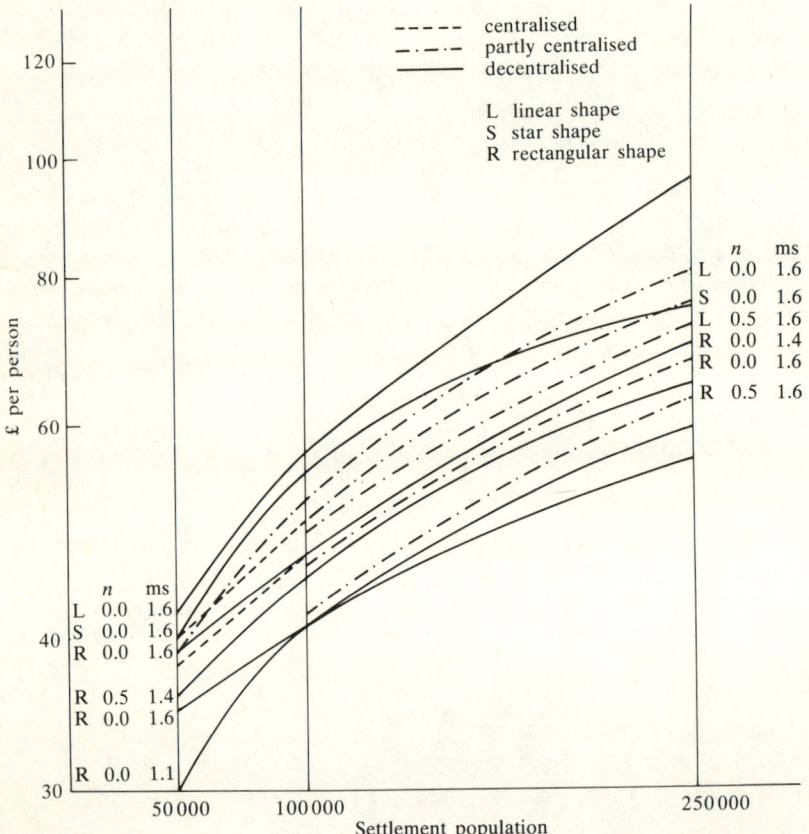

Figure 3.3. Travelling costs and settlement factors (costs are at 1967 prices; tanner n is a constant in the gravity model, and modal split, ms, is an index relating to the proportion of people travelling by different modes of transport) (source: Stone, 1973).

more than proportionately with population size. Average travelling costs per head were 22% lower in settlements for 50 000 than in those for 100 000 and nearly 50% higher in settlements for 250 000 (Stone, 1973) (figure 3.3). Where distance had a fairly strong deterrent effect, savings in travelling costs could be achieved by using clusters rather than single large settlements of 250 000 people. Caution must be exercised in interpreting these results in terms of energy requirements, since travelling costs in the model were based on total vehicle costs (including fuel) and travel time. It is also apparent that the 'efficiency' of a particular spatial structure depends on the values assumed by 'nonspatial' variables such as the deterrent effect of distance. A cluster of smaller settlements may offer greater *potential* for energy efficiency than a single large one. But if people are not deterred by distance, they may travel freely between settlements to increase the range of opportunities open to them. In at least one other theoretical study it has been suggested that per capita transport energy consumption would be greater in a spatial pattern of many small settlements than in one with a small number of large centres, reflecting individuals' need for access to facilities which some 'villages' would lack (even though it was assumed that people would work in their own settlements) (Fels and Munson, 1975).

Contradictory findings also come from work in the USA. In one regression analysis based on data from 134 metropolitan areas, a significant *negative* correlation between per capita fuel consumption (for private transport) and total population was found (Stewart and Bennett, 1975). But Keyes and Peterson (1977), using data from 49 metropolitan areas, found quite the opposite; and, in contrast to findings in the United Kingdom, positive correlations of average *trip length* with population size have also been demonstrated (Bellomo et al, 1970; Vorhees, 1968).

These apparently contradictory results are indicative of the number and complexity of factors influencing transport energy consumption. Levels of car ownership, availability of public transport, and spatial separation of activities all vary between the types of area considered. The relatively high energy use in Greater London, for example, in spite of the important contribution of public transport services there, is attributed to high levels of car ownership (Maltby et al, 1978). The increase in average trip length with urban size in the USA may well be due to urban sprawl and the resulting physical separation of activities. It is clearly difficult to isolate the effects of urban size, and apparent correlations may be due to the operation of other factors which may not always be related to urban size in the same way. Important variables influencing the relationship between size and energy consumption certainly include the range of facilities offered in any particular settlement and the deterrent effect of distance. Ceteris paribus, travel requirements would probably be lowest in relatively small self-sufficient settlements whose inhabitants were content to use the facilities available locally. But this spatial pattern may be energy inefficient when travel is uninhibited.

Shape
The question of an ideal shape has also intrigued urban planners, but again it has an academic air about it, since there are so few opportunities to plan a new settlement on a clean slate. The question has not been answered to anyone's satisfaction—indeed, it has not even been addressed in a consistent form, since 'shape' is a rather elusive concept, usually defined in terms of the transport network and embodying a large variety of assumptions and criteria for evaluation. There is a general consensus that circular settlements are inefficient in terms of transport and energy requirements, and that linear or rectangular forms have some advantages, at least in theory (Edwards and Schofer, 1975; Jamieson et al, 1967; Stone, 1973). But the forms investigated have been highly abstract, and it seems probable that travel needs and transport energy use will depend less on the overall 'shape' of a settlement, defined in terms of its transport network, than on the internal arrangement of activities and, of course, on nonspatial factors. We now turn, therefore, to the intraurban scale, where much more of immediate practical interest can be said about energy–land-use interactions.

The intraurban scale
At the intraurban scale there is more consensus about energy-efficient spatial characteristics, since most empirical and theoretical evidence points in the same direction. The single most important factor in the relationship between urban form and transport energy requirements seems to be the physical separation of activities, determined in part by density and in part by the interspersion of different land uses. The lower the physical separation, the lower travel needs are likely to be, though an 'efficient' spatial structure alone will not guarantee transport energy savings.

Density
Density has perhaps received more attention than any other single variable in the energy–urban form debate. This emphasis may well be responsible for the notion that energy efficiency requires futuristic 'compact cities'—a notion which can be easily dispelled. It is certainly true that density tends to be negatively correlated with transport energy use, but it is not the case that energy efficiency can be achieved only with very high densities, nor should high density be confused with 'high-rise' development.

Having reviewed a number of empirical studies on density and travel or transport energy requirements in the USA, Keyes (1982, page 220)

concludes rather cautiously that
> "Otherwise comparable households who live in low density urban areas or on the periphery of large metropolitan areas tend to travel further, faster and much more frequently by auto than their counterparts in high density cities".

This conclusion is supported by cross-sectional analyses of different metropolitan areas, different counties (for example, the thirty-one counties of the New York City Region), and different areas within the metropolis [for example, see Keyes and Peterson, 1977; Neels et al, 1977 (discussed by Keyes, 1982); RPA, 1974]. Differences seem to be explained less by the number of times people use their cars, than by how far they travel once they are in them, pointing to the influence of the physical separation of activities (reduced by high densities) as an important factor. The inverse relationship between urban density and journey length is also found in the United Kingdom; work trips increase in length from an average of 1.37 miles in North Central London to an average of 5.6 miles in Kent (GLC, 1971; Hamer, 1976). Theoretical comparisons between different urban forms, using urban models to estimate travel, also quite consistently support the inverse relationship between densities and transport energy consumption (Clark, 1974; Edwards and Schofer, 1975; RERC, 1974; Roberts, 1975).

Interspersion of activities
Compact development is not the only way of reducing the physical separation of activities; this can also be effected by interspersion of residential and other land uses, and by 'clustering' employment and services (see figure 3.4). This implies that some decentralisation of the latter two might be desirable in order to achieve "more effective integration at a smaller geographic scale" (Odell, 1975, page 48).

It can be readily demonstrated in modelling exercises that dispersal of employment and service opportunities could be energy efficient. Highly abstracted models of urban form explored by Hemmens (1967) (in perhaps the first study of this kind) and by Schneider and Beck (1973)

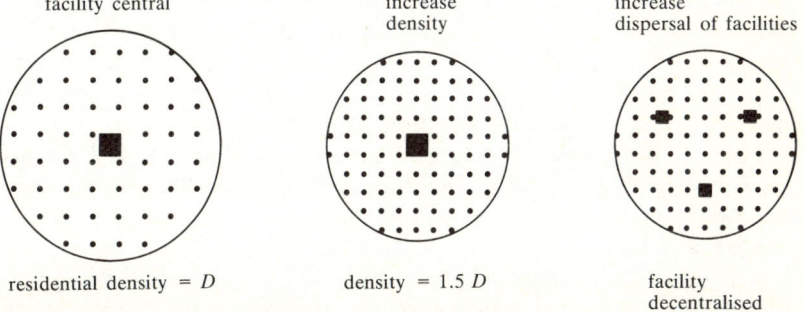

Figure 3.4. Decreasing physical separation of activities.

strongly suggest that decentralisation of jobs and services would save energy—in a model of King County, Washington, explored by Schneider and Beck, travel requirements could be more than halved by moving between a half and a third of the jobs.

These models suggest a rather deceptive simplicity, however. When factors like the propensity to travel, the premium placed on *choice* of jobs and services, and the extent to which cars are used are introduced, it becomes obvious that we cannot place land-use patterns in any absolute order of energy efficiency—their relative performance depends on the values assumed for these nonspatial variables. In a study of twelve alternative growth patterns in rural Norfolk, for example, in which the physical separation of homes and jobs varied by up to 350%, it was found that the *ordering* of alternative growth patterns in terms of energy requirements for the journey to work depended on assumptions made about the deterrent effect of distance (Owens, 1981). When this was high, the most efficient growth patterns were those which led to low physical separation of homes and jobs. (In the particular area studied, this involved concentration of new population into the main town but maximum dispersal of new employment.) Physical proximity did not have such a marked influence on energy requirements when a low deterrent effect of distance was assumed (with a maximum difference in energy requirements of 41% between alternative growth patterns, as compared with 128% with high deterrence). In this case the most efficient pattern involved concentration of both population and employment growth. In other studies it has also been found that centralisation would result in lower travelling costs if distance was *not* a deterrent, but in higher costs if it was (for example, Stone, 1973).

The important implication of these findings is that it may not be possible to identify 'intrinsically' energy-efficient land-use patterns. A particular pattern may be efficient when there is low mobility, but quite energy-intensive when distance is not a deterrent. The reduction of the physical separation of activities is not a sufficient condition for the reduction of transport energy requirements; it must be accompanied by a decreased *inclination* to travel longer distances and, to some extent also, a willingness to take advantage of the opportunities offered for nonmotorised travel. Furthermore, if people are not going to be very much deterred by distance, energy requirements may actually be minimised by land-use patterns which do *not* minimise physical separation. It is not difficult to find examples of areas which were intended to be relatively self-contained but which continue to experience considerable in- and out-commuting. This has been observed in new communities where positive attempts have been made to relate residential and other land uses (Keyes, 1976), and also in relation to incremental development. A policy in the Structure Plan for West Sussex, seeking to locate new development in ways which would minimise travel, has had 'little practical effect' and is likely to be deleted when the Plan is reviewed (P W Bryant, personal communication, 12 December 1983) (though one wonders whether this evaluation is rather premature).

The problem is that planners will not know the future values of the variables which influence the relative efficiency of land-use patterns, apparently introducing a major element of uncertainty into 'energy-integrated' planning. But it is also possible to identify 'robust' options which perform relatively well under a range of possible future conditions: in the Norfolk study, for example, a growth pattern involving concentration of new population and some degree of dispersal of new employment was judged to be the most robust (Owens, 1981).

'Clustering' of land uses is also likely to reduce travel needs, by reducing physical separation and making multipurpose trips more feasible. This concept is not in contradiction with that of interspersion of land uses since it is, of course, quite possible to have 'dispersed clusters'. Savings might be quite considerable. Analysis of travel diaries collected in the New York Metropolitan Area, for example, suggested that clustering of commercial and industrial establishments resulted in a reduction of 65% in trips per thousand square feet of floor space as compared with dispersed facilities (Markovitz, 1971).

Summary so far
Comparative analysis of different structures suggests that an energy-efficient settlement pattern would consist of small- to medium-sized settlement clusters. Within settlements, overcentralisation of employment and services would be avoided. Instead, residential areas would be planned around more dispersed clusters of employment and services in relatively compact 'urban subunits'.

It is important to realise that high-rise development would not necessarily be a feature of this settlement pattern. Quite moderate densities could achieve all of the objectives. With twenty-five dwellings per hectare, for example, facilities with a catchment area of 8000 people would be within 600 metres of all homes (Dickins, 1975), and a 'pedestrian scale cluster' of 20000-30000 people could provide a sufficient threshold for many facilities without resort to high densities (Thomas and Potter, 1977). This again points to the importance of the interspersion of land uses. In fact, there is evidence to suggest that very high densities may be counterproductive in energy terms. 'High-rise' developments require the use of highly energy-intensive construction materials, greatly increasing the 'energy investment' in infrastructure. They may have high heating requirements (due to weather exposure) and will almost certainly be inimical to the use of dispersed renewable energy. The construction and operating costs of high-density residential development may be a half to two thirds greater than those associated with medium density (Stone, 1973), and although these figures refer to financial costs, there are clear implications for the energy investment in the construction and running of urban areas.

Promoting energy-efficient transport
So far, the discussion has focused mainly on ways in which the need to travel might be reduced. It is also frequently argued that a shift towards public and nonmotorised transport from energy-greedy private vehicles would reduce fuel consumption, as well as conferring social and environmental benefits which are already widely recognised. Many of the structure plans produced by local authorities in England and Wales, for example, include policies to 'promote public transport', justified by one or more of these objectives. It is also clear that certain land-use patterns are much better suited than others to the efficient and economic operation of public transport; hence the intuitively attractive argument that appropriate land-use planning policies could contribute to energy conservation by providing the right conditions for an energy-efficient modal shift in transport.

This argument takes us into the intensely political realm of transport policy, and particularly into the wide-ranging, often heated, and continuing debate about public versus private transport, and its strong ideological crosscurrents and conflicts of interest among the 'motor lobby', transport operators and users, and environmentalists. Important issues in this debate are therefore touched upon in what follows. Nor is it forgotten that there is in theory considerable potential to influence energy demand in the transport sector directly, rather than through its relationship with land use. Banister (1981) has provided an excellent summary of the scope for energy conservation in this context and has identified certain specific policy areas where local authorities would be involved (sometimes in conjunction with central government). These include promotion of public transport (which may involve politically controversial subsidies), provision of pedestrian and cycling facilities (which does have some land-use implications), consolidation of freight loads, urban traffic control, parking control, and area licencing. He also considers the potential for car sharing, but suggests that at the local level, most benefits come indirectly from traffic management schemes. The point to be emphasised is that 'nonspatial' policies, as well as measures aimed at modifying the form of the built environment, may be effective in reducing transport energy requirements. Reference will be made to some of these policy options, but emphasis will be on those aspects which are fairly directly related to land-use patterns.

Public transport is potentially more energy efficient than private transport, with rail systems occupying a position somewhere between that of buses and that of private cars. Whether it is *actually* more efficient depends on load factors, so that it has proved surprisingly difficult to specify energy-consumption performances for different travel modes.

On the basis of *average* passenger loads, a bus travelling in uncongested road conditions is the most energy-efficient, and a car used for commuting in an urban area is the least (Maltby et al, 1978). In urban areas, a bus may be four times more energy efficient than a car, but at off-peak periods and in rural areas, occupancy levels are low, and it is often pointed out that with higher car occupancy, the energy advantages would be less clear. There is an inconsistency in this argument, however, since one reason that occupancy levels are low is that public transport is forced to compete with private vehicles in a highly imperfect market. In any case, at present, with the prevailing one or two passengers per car, it seems safe to say that public transport is actually, as well as potentially, more energy efficient than travel by private vehicle, although its full advantages are not realised.

From an energy conservation point of view then, a shift from private to public transport (or of course to walking and cycling) is desirable. Savings could be significant, but not enormous, as shown by the figures for work and other journeys, from estimates by Maltby et al (1978), in table 3.1. Even modest savings may be worth making however, and other substantial benefits might accrue from such a policy, in terms of environmental quality and accessibility for the 'transport poor' (for example, see Adams, 1981; Hillman et al, 1973).

Modifying spatial structure is one measure which might be used to promote public transport (others include fare subsidies, traffic management, and restrictions on the use of private vehicles). As noted, some land-use patterns are more conducive to public transport than others. Relative concentration of homes and facilities maximises accessibility to the transport route and encourages a high load factor (Jamieson et al, 1967; Roberts, 1975), whereas public transport is particularly bad at serving dispersed low-density areas typical of residential suburbs. There is some empirical support for this view. For example, in Keyes and Peterson's (1977) analysis of data from forty-nine metropolitan areas in the USA, in which they found an inverse relationship between density and per capita petrol consumption, the effect of the 'high density' variable was weakened when the extent of the public transport system

Table 3.1. Energy savings from modal shift (source: figures given in Maltby et al, 1978).

Modal shift	Potential energy savings
Transfer of 50% of urban work trips by private vehicle to bus services	8 – 12%
Reduction by 50% of shopping and personal business travel by private vehicle	3 – 6%
Reduction by 50% of other personal and social travel by private vehicle	16 – 19%

was considered, suggesting that high density acts by improving access to transport, as well as by clustering trip ends.

The implication of all this is that planners should encourage higher residential densities and concentration rather than dispersal of facilities. But, although it is undeniable that increased densities in general favour public transport, there are also significant *shape* factors involved at the intraurban scale. For example, a 'linear grid' form could combine high linear densities with moderate overall densities along the route served, compatible with a high-quality environment (Steadman, 1980). Relevant planning policies might therefore include discouragement of dispersed low-density suburbs, some degree of clustering (though not necessarily *centralisation*) of facilities, siting new development near to transport routes, and the maintenance of moderately high densities *along* these routes. Such policies could be applied to green field developments and also to redevelopment, growth, and 'infilling' in existing settlements.

As with the reduction of travel needs, however, appropriate spatial structures provide a necessary but not a sufficient condition for the successful operation of public transport—a simple caveat which is all too often overlooked in the enthusiasm for 'ideal' forms. The viability of public transport depends on many interrelated factors, including car ownership, socioeconomic characteristics of the population, past investment in infrastructure, and public policy towards transport in the area concerned.

Public transport has certainly experienced severe and accelerating decline in recent years. In the United Kingdom, adult patronage decreased by an average of 30% between the mid-1970s and the early 1980s. Increasingly it is recognised that this decline is rooted in the inherent inability of public transport to compete with private transport in a system in which external costs and benefits go largely unaccounted for. In the absence of any significant restraint on car use, a 'vicious circle of decline' is established (figure 3.5)—driven essentially by rising incomes—which cannot be reversed by fare reductions, traffic management, or indeed appropriate urban development policies (Hills, 1983).

There is a fairly considerable body of evidence to support this view at least in the British context. Traffic management schemes, such as bus priority systems, have not been notably successful (see Heggie, 1977). Lower fares increase the ridership of public transport, but mainly by 'captive' users. Although this represents a social benefit which should not be lightly dismissed, success in attracting motorists back out of their cars has been limited. The effects of subsidies in encouraging transfer from car to bus or rail seem to be rapidly offset by increases in income; one model suggests that for every 1% increase in income, a 3% increase in subsidies would be required to maintain ridership (Oldfield et al, 1981). Policies which seek to restrain the use of the private car (such as road pricing or control over private parking spaces) have typically foundered on the rock of effective political opposition from vested interests (for example, see Banister, 1984; Starkie, 1982). Work continues on these effects and no doubt further measures will be attempted.

There are also some isolated exceptions to the general decline, such as the attractive and successful Tyneside 'Metro'; and some measures, such as the low-fares policy in Sheffield, need a longer period of evaluation. But in general, public transport continues to serve a 'captive' market and has an increasingly 'welfare' image (Hills, 1983). It is difficult to see why modification of land-use patterns should succeed where other measures have so obviously failed. Indeed, in the absence of greater control over the factors influencing public transport, policies designed to save energy may even have quite the opposite effect to that intended. In Devon, development was encouraged in some remote communities in an attempt to safeguard rural transport services; but the decline in transport has proved to be beyond the influence of such policies, so that the effect has actually been to make even more people dependent upon private transport (Owens, 1986).

Since the unplanned outcome of the vicious spiral is undesirable in terms of energy, economy, equity, and environment, the only option is to grasp the nettle of more severe restraint on private vehicles. Hills (1983), for example, proposes a radical 'manifesto' which includes suspending the rules of competition for certain routes, imposing severe

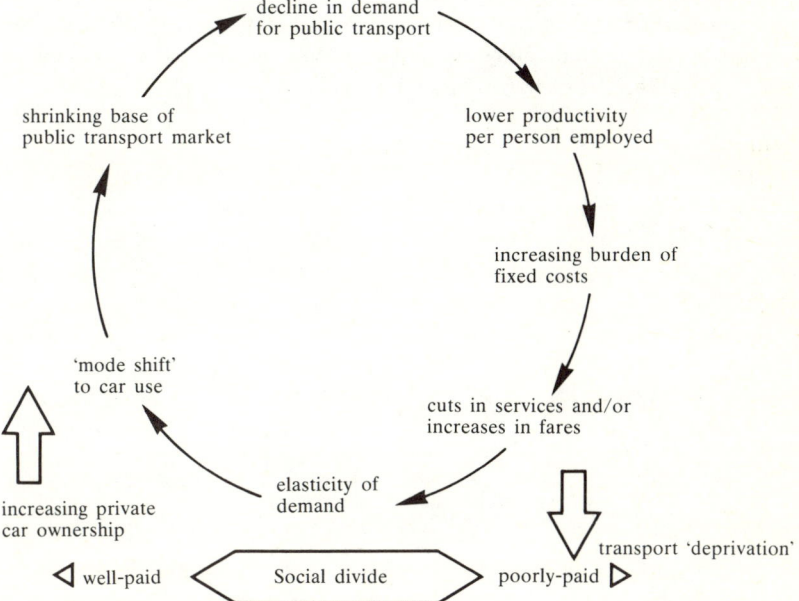

Figure 3.5. The vicious circle of decline of public transport (source: Hills, 1983).

restraint on parking and entry by car into some areas, and discriminating heavily and consistently in favour of public transport in central areas. Current transport policy is almost the exact opposite, with increasing emphasis on competition and the notion that public transport should 'pay its way'. [This philosophy is evident, for example, in the Transport Act (1983) and the London Regional Transport Act (1984). See also the White Paper on Bus Policy (Department of Transport, 1984).] If the above analysis is correct and unless there is a significant shift in thinking at the policy level (which is itself not implausible—there have been many shifts in policy emphasis in the past few decades), then the outlook for public transport is bleak. Perhaps Adams (1981) puts his finger on the real issue when he argues that there is a fundamental inconsistency in the need for an equitable and environmentally sound transport policy in a country which also seeks to maintain and promote a viable motor industry. Certainly, few national or local policies which have sought, however timidly, to restrict the freedom of motorists have survived the intensive political lobbying which they inevitably arouse.

Once again what emerges is a need to fill the void between theoretical possibility and practical reality. Land-use planning which encouraged higher densities, especially along transport routes, and clustering of employment and services could make a modest but significant contribution to the more efficient use of energy resources. But there are formidable political barriers to be overcome in transport policy before this contribution can begin to be made. Significantly, constraints arise more from powerful vested interests than from any real conflict between energy and social or environmental objectives.

Reducing energy requirements in buildings

Introduction
Buildings form the other important link in the energy-spatial structure relationship. In the United Kingdom, for example, building services account for at least 40%, and probably as much as half, of total primary energy requirements, largely for space (and water) heating (BRE, 1975). This aspect of the relationship is manifested mainly at the local scale, where the responsibilities of the planner overlap with those of the architect and builder. The 'fine grain' of urban development can have a very marked effect on the energy budget and deserves much greater attention than it currently receives in Britain.

Built form is one very significant variable at this scale, since the basic energy requirements of a building are determined by its surface area: volume ratio. Siting and orientation have important energy implications since they can be used to gain advantage from microclimatic factors and from 'free' ambient energy sources. Orientation, layout, and density also facilitate or prejudice the introduction of district heating networks. In short, spatial structure at the local scale both influences energy demand and to some extent dictates which energy-conserving technologies are feasible.

In this chapter, first those built forms which are intrinsically energy-efficient are identified, then the spatial constraints associated with renewable energy resources and with combined heat and power generation (CHP) and district heating (DH) systems are considered. Clearly, since infrastructure is so long lasting, it is important to identify these constraints if options are to be kept open for the future. And some important questions need to be answered. Can an energy-efficient spatial structure be defined in the absence of a firm commitment to any particular energy-conserving technology? Are the spatial requirements of different technologies mutually exclusive, or is it possible to define a flexible and efficient spatial structure which is compatible with a variety of measures for energy conservation? Would an energy-efficient environment mean great sacrifices in terms of amenity, equity, or aesthetics?

Other important questions which deserve some attention relate to broader and less well-defined issues, including the spatial implications of extensive use of renewable resources (decentralisation?) and the possible influence of large-scale investment in CHP/DH on urban trends.

Energy efficiency and built form
Theoretical considerations show clearly that if other variables are held constant, built form exerts a systematic influence on energy requirements for space heating. Since the heating requirements (that is, the *useful* energy requirements) of a building may be calculated quite simply from

basic information about its size, shape, and structure, internal and external temperatures, and the ventilation rate, the relationship can easily be explored.

In a much quoted study, the British Building Research Establishment (BRE, 1975) compared the heating requirements of hypothetical dwellings of similar volume and insulation standards, but of different form. The results (figure 4.1), confirmed by others (for example, Barnes and Rankin, 1975; Granum, 1976), show that a detached house could have energy requirements for space heating three times greater than those of an intermediate flat of equivalent size. This difference is of similar magnitude to that between poorly insulated dwellings and those with medium insulation standards, implying that "any widespread trend in the built form of new housing could have an important influence on the national consumption of energy, in extreme cases of the order of magnitude envisaged from a general improvement in the thermal insulation of new constructions" (BRE, 1975, page 20).

It is much more difficult to demonstrate a clear empirical relationship between built form and heating requirements because of the number of variables involved. Studies based on a large variety of sources (including published statistics, records of public utilities, household surveys, and, in some cases, individual metering of dwellings) all show enormous variation in domestic energy requirements. Loudon and Cornish (1975), for example, found a fivefold variation in annual fuel consumption in similar local authority houses in Scotland. Apart from the obvious influence of climate, other important determinants of household energy use are socioeconomic factors, especially income (Commoner, 1971; Gray, 1955; Halvorsen, 1974), as are behavioural

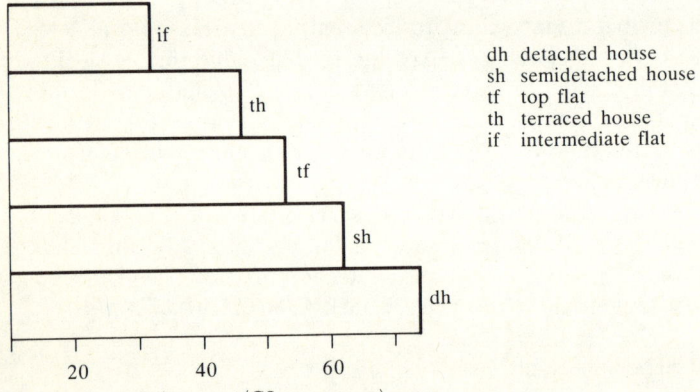

Figure 4.1. Influence of built form on heating requirements (source: derived from Crown copyright BRE information—see BRE, 1975).

variables, such as temperature preferences, daily life patterns, and even window opening habits (Grot and Socolow, 1973).

Empirical work both in the USA and in the United Kingdom has demonstrated significant correlations between domestic energy consumption and the percentage of various types of dwellings in any given area [for example, see Anderson (1973) and Lokmanhekin and Harvey (1974) for the Baltimore-Washington region, Keyes (1976) for New York, and Ball et al (1981) for London]. The problem with such studies is that although they may indicate that energy demand is related to spatial structure, it is not usually shown how much of the variation may be attributed to built form alone as opposed to other variables such as the size of the dwelling and the income of its occupants. In the Baltimore-Washington region of the USA, for example, detached houses consume on average more than twice as much energy as high-rise flats, but differences in energy requirements per square metre of floor space or per occupant are less marked, as shown in table 4.1. Clearly, space standards between different types of dwelling account for a large part of the variation.

The argument in relation to empirical studies tends to become rather circular. Because of the very large number of variables influencing energy use, it is difficult to isolate the effects of built form. A considerable effort has to be made (in terms of data collection and analysis) in order to do this, and when it is done (if it can be) it tends to show that less energy is required by built forms with lower surface area : volume ratios—which we know *must* be the case, ceteris paribus. In other words, a complex analysis is conducted to prove something which is already known to be true. The value of such studies must lie in the ability to indicate (if carefully conducted) the *relative* significance of built form as opposed to all the other factors exerting an influence on energy demand. The indications are that this is rather small.

It can be concluded that the energy advantages of certain built forms—notably terraced houses and low-rise flats—are now well recognised. As yet, these advantages seem to have little influence in the housing market,

Table 4.1. Average annual net energy requirement for heating and cooling typical housing units in the Baltimore-Washington area (source: Keyes, 1976).

Structure	Energy requirements (GJ per annum)		
	per unit	per m^2	per occupant
Detached house	105-116	0.67-0.73	26-29
Town house	94	0.78	24
Low-rise flat	61	0.58	20
High-rise flat	52	0.57	26

or on the proportions of different kinds of new dwellings built—other factors are still much more important. Arguably, energy conservation can be achieved here only at the cost of some loss of amenity. This is not necessarily the case with the possibilities for making better use of ambient energy sources, considered in the following section.

Renewable energy and spatial structure
The local scale
While relying to a greater or lesser degree on conventional energy sources, buildings and settlements can be planned in order to take maximum advantage of the energy cycles of the natural world. Siting, layout, and orientation of buildings determine the extent to which they can be warmed by or shaded from the sun, and ventilated by or protected from the wind. Landscaping, for example, the use of trees to provide shade in summer, is also significant at this scale.

Traditional built form and microlocation in different climatic regions show clearly that the benefits to be obtained from siting in relation to microclimate have long been realised in practice. Piute Indians, for example, skillfully used microclimatic differences in Owens Valley, California, in locating their mobile communities, moving their camps to locations on the east or west slopes to maximise or minimise their exposure to the sun according to the season (Knowles, 1974). Indians in hot arid zones built 'multifamily' homes, sharing thick adobe walls, to protect them from extreme heat and glare, usually oriented on an east–west axis to minimise the direct rays of the sun in summer and to maximise them in winter (Harwood, 1977). At the opposite extreme, Eskimo igloos in the Arctic deflect the wind and take advantage of the insulating effects of snow. There are many more examples [Steadman (1975) provides a very useful account]. In the past, human beings have almost instinctively adopted the principle of designing to take advantage of the natural environment; it is more recently that we seem to have forgotten, or failed to adapt, this concept and have attempted to maintain levels of comfort (not always successfully) through the consumption of nonrenewable fuels.

Now that energy conservation has become a significant issue, interest in siting in relation to microclimate has been reawakened. Some widely accepted criteria are shown in table 4.2. There is particular interest in making use of solar energy for space and water heating, and the requirements of these systems in cool and temperate climates provide an excellent example of the energy–spatial structure relationship at the local scale.

Table 4.2. Site orientation chart (source: Keplinger, 1978).

Adaptations	Objectives			
	cool regions: maximise warming effects of solar radiation, reduce impact of winter wind, avoid local climatic cold pockets	temperate regions: maximise warming effects of sun in winter, maximise shade in summer, reduce impact of winter wind but allow air circulation in summer	hot humid regions: maximise shade, maximise wind	hot arid regions: maximise shade late morning and all afternoon, maximise humidity, maximise air movement in summer
Position on slope	low for wind shelter	middle-upper for solar radiation exposure	high for wind	low for cool air flow
Orientation on slope	south to southeast	south to southeast	south	east southeast for afternoon shade
Relation to water	near large body of water	close to water, but avoid coastal fog	near any water	on lee side of water
Preferred winds	sheltered from north and west	avoid continental cold winds	sheltered from north	exposed to prevailing winds
Clustering	around sun pockets	around a common sunny terrace	open to wind	along east–west axis, for shade and wind
Building orientation	southeast	south to southeast	south towards prevailing wind	south
Tree forms	deciduous trees near building evergreens for windbreaks	deciduous trees nearby on west no evergreens near on south	high canopy trees deciduous trees near building	trees overhanging roof if possible
Road orientation	crosswise to winter wind	crosswise to winter wind	broad channel east–west axis	narrow east–west axis
Materials coloration	medium to dark	medium	light, especially for roof	light on exposed surfaces dark to avoid reflection

Spatial requirements for passive solar energy
A cost–benefit matrix for passive solar energy (that is, solar energy harnessed by architectural means without the aid of mechanical or electrical equipment) is shown in figure 4.2. Considerable energy savings can result from relatively simple measures, which do not necessarily incur additional economic or environmental costs. But there are some spatial constraints.

The main requirement for taking maximum advantage of passive solar energy is that buildings should have a north–south orientation with an elongated south elevation. The simplest adaptation is then to have rather larger windows on the south side and smaller ones to the north. Conservatories may also be used. However, there is obviously a need to avoid overshadowing, and this has rather broader planning implications for the layout of neighbourhoods and for development density.

Spacing requirements for passive solar energy are generally more stringent than those which meet other typical planning conditions. As a 'rule of thumb', with two-storey terraced houses on a flat site, daylighting criteria would require 6–10 metres between rows, sunlighting criteria 10–16 metres, 'privacy' 15–18 metres, and use of passive solar energy 20–23 metres—even this spacing would produce some shading during winter (Turrent et al, 1981).

The orientation requirements and the need to avoid overshadowing might seem to imply that the use of passive solar energy would be compatible only with relatively low housing densities. Taking the simple case of terraced housing again, there are clearly physical limits to density above which overshadowing by adjacent rows would present a serious problem. However, recent research suggests that there would not be difficulties below densities of about 30 dwellings per hectare, and even at densities of around 40 dwellings per hectare, loss of solar radiation need not be more than 20% (Ó'Catháin and Jessop, 1978). With design ingenuity, passive solar energy could be compatible with densities approaching 50 dwellings per hectare (Turrent et al, 1981). Since even 30 dwellings per hectare is not a very low density, it may be concluded that the use of passive solar energy would not demand unrealistically low dwelling densities, as has sometimes been implied. Nor are the orientation criteria very rigid; there is a good tolerance limit of 30–40° variation from a north–south axis within which advantage can be taken of solar gain (Doggart, 1979; Turrent et al, 1981).

The absence of severe constraints on density and orientation forestalls several potentially important criticisms. One is that design to exploit passive solar energy might result in rigid repetitive layouts with little aesthetic appeal. A second is that housing patterns for passive solar heating would be 'exclusionary', available only to those who could afford low-density housing on large plots. A third is that energy savings associated with higher densities would not be realised.

Reducing energy requirements in buildings

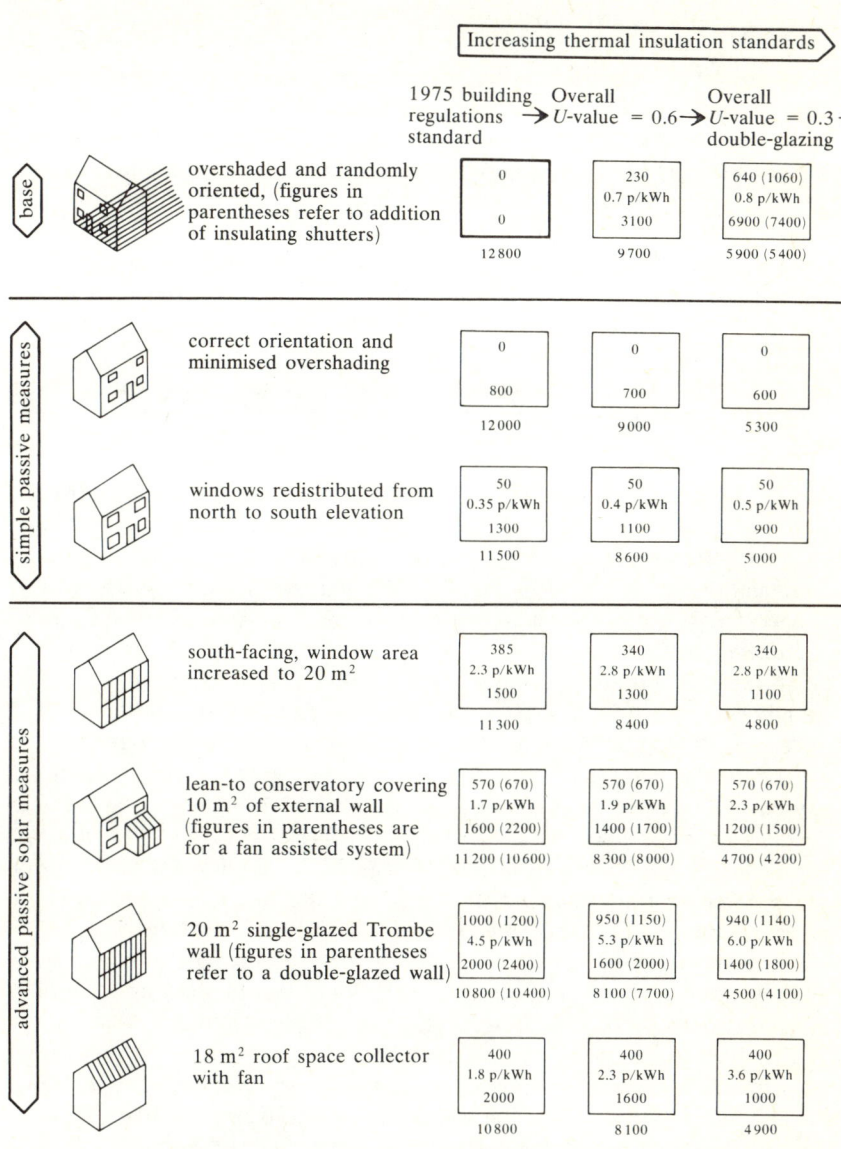

Figure 4.2. Cost-benefit matrix for new housing incorporating passive solar measures. [The main technical options for passive solar heating are summarised. In each box the top figure refers to the extra cost (£) and the bottom figure to the energy saving (kWh per year). Figures below the boxes refer to the residual space-heating loads (kWh). p/kWh is the effective cost of energy assuming a capital repayment factor of 0.09. £270 credit is allowed for value of extra space.] (source: Turrent et al, 1981)

Other physical constraints may, however, be important, particularly the size and shape of the site. As sites become smaller, opportunities for selective orientation decrease. In urban areas, 'infill' sites are often small (typically in the range of 0.05-0.20 hectares), so that factors such as height and proximity of adjacent buildings, access, and drainage become the prime determinants of layout (Turrent et al, 1981). With green field sites, there is more scope to choose the most appropriate orientation and spacing of housing blocks. This means, of course, that the exploitation of passive solar energy could be facilitated by developing new residential areas on the urban fringe or beyond, rather than 'infilling' within existing city boundaries. For this reason, and not because it imposes density constraints, it might be argued that the exploitation of passive solar energy might be associated with 'urban sprawl'.

Spatial structure and active solar systems
The use of 'active' solar power, that is, using mechanical or electrical equipment with a greater or lesser degree of sophistication, may have rather different implications for spatial structure from those of 'passive' applications; this will depend to a large extent on the proportion of energy requirements intended to be met and the scale on which the source is exploited.

The most promising application for active solar energy on a relatively small scale is in the provision of low-grade heat for space and water heating. The economics of this technology depend on the particular location, the pattern of heat demand, and the technical details of the system. In general in the USA and the United Kingdom, solar heating costs seem to compare favourably with electricity, but cannot yet compete with oil and gas (Eden et al, 1981, page 193). Economics may be more favourable in many parts of the world and are likely to improve with technical innovation and as economies of scale are realised.

Heat is typically provided through a roof-mounted flat-plate collector, through which water is circulated. This may be plumbed into the hot water system of the building. All space heating and hot water requirements can be met, but collector and storage requirements drop considerably if there is a 'backup' system to provide, say, 25% of the heat. For roof-mounted solar collectors of this kind, meeting 50% or more of the total heat requirements of a dwelling, orientation and overshadowing are unlikely to present problems; the spatal requirements are less demanding than those for houses making maximum use of passive solar energy (MKDC, 1982).

Spatial implications become more significant if we consider meeting a higher proportion of a community's energy needs with solar power. If more than low-grade heating requirements were to be met, some solar energy would have to be used to generate electricity. This can be achieved by thermal-electric means—solar energy is used to heat a fluid which then generates electricity through a conventional heat engine—or by the use of photovoltaic cells, which employ the properties of semiconductors such as silicon. Large areas are then required to harness solar energy.

Taking an extreme case of a mixed residential, commercial, and industrial community designed to rely exclusively on solar thermal and electric energy for stationary energy needs, it has been estimated that an area of at least 40% of the total land area of the community would be required for collector surfaces (Owen Carroll and Udell, 1982). The space occupied by the average household becomes insufficient to meet collector needs at densities much above 23 dwellings per hectare. This would confine such systems to low-density suburbs in the USA and would virtually rule them out in the United Kingdom. It implies that total reliance on renewable energy sources is not compatible with highly urbanised spatial structures. However, widespread use of solar-generated electricity still depends on changing relative economics and on technical breakthrough, and total reliance on solar systems remains a fairly remote possibility. It can be concluded that the most likely application of solar technology in the foreseeable future, for the provision of low-grade heat at the local scale, need not have very exacting spatial requirements.

Wider spatial implications of 'soft energy paths'
At the local scale the potential for exploitation of ambient energy sources is directly influenced by structural variables related to siting, layout, orientation, and density. These considerations are largely technical in nature. But the suggestion that renewable energy sources may be incompatible with 'urbanisation' raises wider issues, less easily defined but perhaps conceptually more intriguing—and certainly worthy of brief mention before returning to the local scale.

Many environmentalists have advocated the adoption of 'soft energy paths' as the only sure way to a sustainable future both for developed and for developing countries (most notably Lovins, 1977). This would entail the use of renewable rather than nonrenewable energy sources and would involve their exploitation on a small and decentralised scale rather than through centralised high technology. Such scenarios inevitably raise quite profound questions about economic growth, distribution, and life-styles when compared with typical projections of world energy demand and the potential contribution of renewable energy sources. For example, work at the International Institute for Applied Systems Analysis (IIASA) suggests that a realistic contribution from renewable energy sources, excluding centralised 'high technology' applications (such as large-scale solar energy), would be of the order of 7 terawatts (TW) (1 TW = 10^{12} watts), against projected world energy consumption in 2030 of 24–40 TW (Anderer et al, 1981; Sassin, 1981).

But apart from the question of their ultimate feasibility in global terms, 'soft' energy technologies also have interesting implications for spatial structure, since concentrated urban agglomerations could not be supported in this way. Urban settlements are likely to generate energy demand densities of 5–10 watts per square metre (w/m^2), perhaps reaching a level an order of magnitude higher than this in specific areas (Sassin, 1981). Comparison of such spatial densities of energy demand with renewable energy flows (figure 4.3) confirms how urban systems 'distort' the natural energy flux by 'borrowing' energy from other times and places, as illustrated in figure 4.4 (Windheim and Wodder, 1976). Although the spatial density of energy demand may well be reduced in future, it seems unlikely that urban settlements could exist on the basis of distributed energy sources and will require at least some degree of centralisation of energy supply systems (Sassin, 1981).

Urban demand densities have been used primarily to draw inferences about future global energy demand based on projected urbanisation trends (for example, Newcombe et al, 1978; Sassin, 1981). But there is also a clear implication that reliance on small- and intermediate-scale renewable energy technologies would require a spatial structure of dispersed, relatively small-scale, settlements. The more immediately practical implications are that urbanised areas must continue to rely

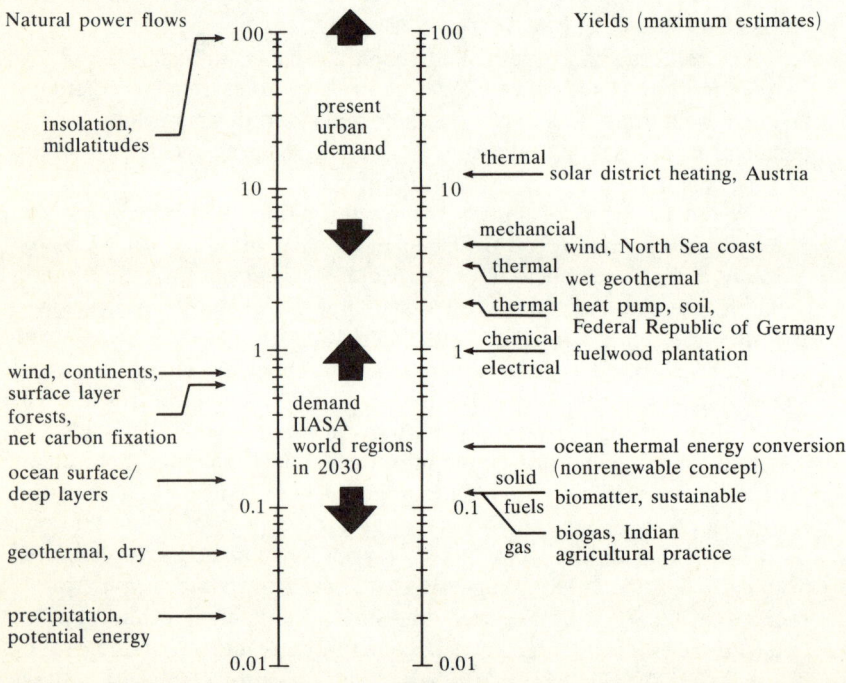

Figure 4.3. Energy supply densities (source: Anderer et al, 1981).

quite heavily on centralised energy sources (though in future, conventional fuels could be replaced by high technology application of renewable energy). Exploitation of ambient energy sources on a small to intermediate scale will be able to meet a proportion of energy demand in such areas, and has considerable potential for decentralised village-scale applications in nonurban areas, particularly in developing countries where there is little existing infrastructure for conventional power supply. [The case for the latter type of application has been strongly argued by Leach (1976).] Although there has been a certain amount of speculation about a European future with dispersed quasi-self-sufficient communities exploiting natural energy systems (*The Ecologist*, 1972; Hall, 1977), such a spatial structure seems at present rather a remote possibility in the industrialised developed nations.

In reality, for the immediate future, it is at the local and intraurban scales that energy demand, the potential to use 'free' solar power, and, spatial structural variables will interact. These are also the levels at which these variables influence the potential for energy conservation through the introduction of combined heat and power generation.

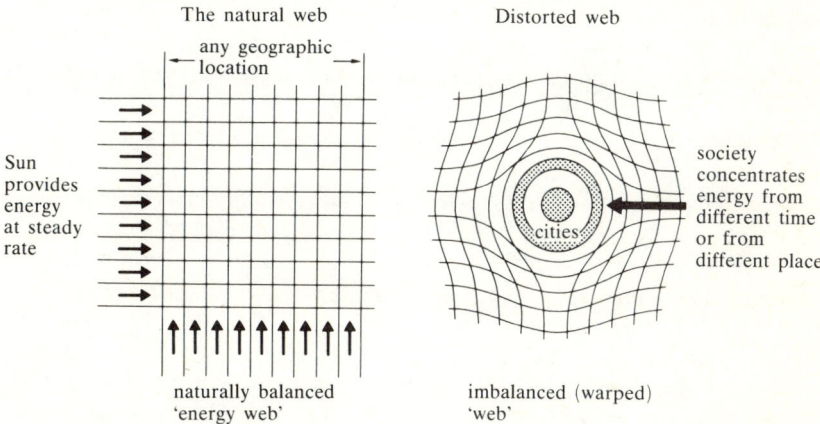

Figure 4.4. How cities distort the natural energy flux (source: after Windheim and Wodder, 1976).

Combined heat and power generation

Conventional power stations convert primary fuel into electricity with a maximum efficiency of about 35%. Some 50–60% of the energy is converted to low-grade heat, usually discharged into the atmosphere or into rivers. The temperature of this low-grade heat is too low for most practical heating purposes. This 'waste' of two-thirds of the primary fuel going into a conventional power station seems particularly indefensible when emphasis elsewhere is being placed on energy efficiency, notwithstanding the defense of the electricity supply industry that it is simply part of the cost of producing higher grade energy, more efficient at the point of use. However, with a particular type of steam turbine it is possible, by reducing the efficiency of electricity generation to about 31%, to produce water at a temperature which is high enough for space heating of buildings (Atkins and Partners, 1982). The production of electricity together with useful heat in this way is known as combined heat and power generation. The overall efficiency of primary energy use with CHP can be as high as 80-90%, representing considerable improvement on the separate production of electricity and heat. Hot water is distributed to buildings through a system of pipes forming a district heating network. DH may also be supplied by heat only boilers, which have efficiency and fuel flexibility advantages over small boilers serving individual buildings, and it is feasible for this to be an intermediate stage before connection of a network to a full CHP scheme.

The energy advantages of CHP/DH are broadly accepted: the main question for those countries considering investing in such systems is whether their economic and social costs can be justified. It is in evaluation of these costs that land-use and planning considerations become significant. The potential for the introduction of CHP/DH depends (among other considerations) on the density of development and on the degree of mixing of different land uses. And there are of course broader planning implications in the sense that any existing area selected for a CHP scheme would be subject to significant impacts (both positive and negative) during the implementation and operation of the scheme.

CHP/DH is already used extensively in Scandinavia and in Eastern Europe, and to a smaller extent in other European countries. In Denmark, for example, 40% of the heat consumption in the residential/commercial sector is provided by DH; the corresponding figure for Sweden is 25%. Interest in this technology in the United Kingdom has grown considerably since the early 1970s. It has been estimated that energy savings from CHP/DH might be as much as thirty million tonnes of coal equivalent per annum, or 5-10% of probable UK energy demand beyond the year 2000 (CHP Group, 1979). The adoption of CHP/DH would also simplify the process of fuel substitution as oil and natural gas become more expensive. A special working group was established by the Secretary of State for Energy in 1974 "to consider the economic role of combined heat and power in the UK and to identify technological, institutional, planning, legal or other obstacles to the fulfilment of that role and to make recommendations" (CHP Group, 1977, page 3).

Spatial requirements for CHP/DH
Theoretical calculations by the UK Working Group (CHP Group, 1977) show how sensitive are the economics of CHP/DH to dwelling density, mainly because of differences in the cost of supplying the hot water network.

Figure 4.5 shows the variation in cost of the network with dwelling density for a 'green field' development and for an existing city.

The Working Group estimated the costs of DH and CHP in three hypothetical situations—a small 'green field' development, an existing small city, and an existing large city—and compared these with the costs of individual gas-fired central heating (for the new development) or of the existing fuel mix. Costs, discounted over a sixty-year period, included those for the network, the boilers or CHP plant, house internals and metering, and fuel.

There is no simple answer to the question of which system is most cost-effective. The relative economics of CHP/DH and the more conventional heating systems depend upon the discount rate, assumptions about real fuel prices, dwelling density, and the type of area being considered. Figure 4.6 (see over) illustrates the effects of variations in dwelling density, discount rate, and fuel price assumptions in the 'small city' case. Costs of transmitting heat from the power station to the scheme (£27 per dwelling for each kilometre) must be added. Assuming a power station 15 kilometres from the edge of the city, a 10% discount rate, and constant real fuel prices, the 'break even' density is more than 250 dwellings per hectare. With a discount rate of 5%, this figure is reduced to about 50 dwellings per hectare. Keeping the 10% discount rate, but assuming that real fuel prices will double every 18 years, also produces a break even density of about 50 dwellings per hectare. In the green field case the break even density is about 75 dwellings per

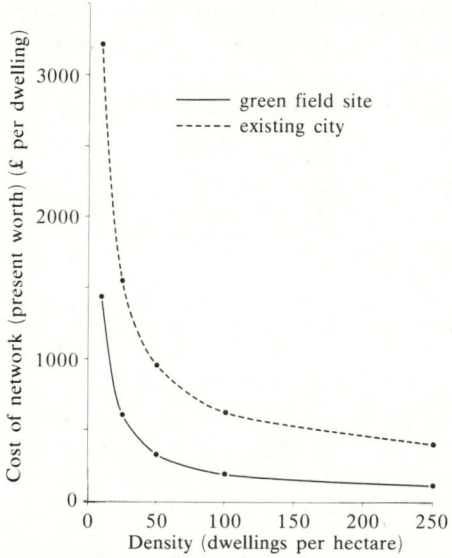

Figure 4.5. Variation in cost of DH network with dwelling density (source: derived from figures given in CHP Group, 1977).

hectare with a 10% discount rate and constant real fuel prices. This can be reduced to about 25 dwellings per hectare or about 37 dwellings per hectare by modifying the discount rate or fuel price assumptions respectively, as above.

The economics of CHP/DH are clearly most favourable in areas of higher density, especially if conversion of existing built-up areas is being considered: but the possibility that it will become economically viable to connect quite large areas to such schemes in the future cannot be ruled out. It is worth noting that further refinement of the models used, in particular to allow spatial variation in the distribution of heat demand, suggests that costs may have been overestimated by the CHP Group (Jebson, 1981). For example, it can be shown that for cities exhibiting a pattern of radially decreasing demand, network costs would be reduced to 60% of those for the less realistic 'uniform' city assumed in the original models, and similar reductions result from assuming a mix of land uses rather than a wholly residential area[4].

In their final report to the Secretary of State, the CHP Group (1979) note a general consensus that fuel prices would double and might even treble by the end of the century, and suggest that doubling might be regarded as a 'reasonable central case'. They also consider that the use of a test discount rate lower than 10% may be justified in circumstances where a successful commercial outcome can clearly be foreseen. They conclude that CHP/DH is unlikely to be viable before the turn of the century, except in very high density areas, mainly because of competition from relatively cheap natural gas; but in the longer term, CHP/DH could provide the cheapest method of heating towns and cities in the United Kingdom. They recommend that a full-scale CHP/DH scheme should be implemented in a lead city.

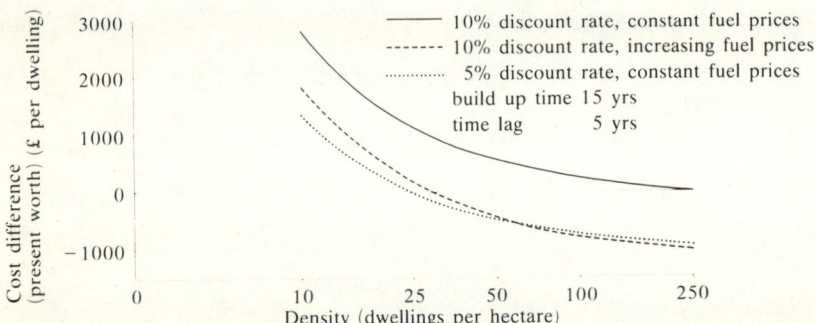

Figure 4.6. Effects of dwelling density, discount rate, and fuel price assumptions on economics of CHP (source: derived from results of computer runs in CHP Group, 1977).

---

[4] In the original model, the heat demand of commercial and institutional properties was included as an additional number of domestic dwellings.

A programme of work on the feasibility of CHP/DH in specific locations was announced by the UK Government in 1980 (see Atkins and Partners, 1982, page i). Consultants were appointed to identify suitable locations with a view to narrowing down the options and to possible selection of a scheme for implementation. Subsequently, detailed studies of the feasibility and economic viability of CHP schemes in nine major British cities [Glasgow, Newcastle, London (Central and East), Sheffield, Belfast, Liverpool, Edinburgh, Leicester, and Manchester] have been conducted, in conjunction with the relevant local authorities and the electricity supply industry.

These more detailed studies of real cities concentrated from the outset on 'high-density head load' (HDHL) areas, defined as those having a net density of 20 megawatts per square kilometre or more, which would be typical of areas with dwelling densities of at least 44 dwellings per hectare. Inner-city areas of this relatively high density typically also contain a mix of residential and commercial floor space which not only provides the large aggregate heat load needed to make most efficient use of the CHP plant, but also makes maximum use of other components of the system, such as arterial mains. Within these areas, results of the studies confirm the inverse relationship between development density and network costs (Atkins and Partners, 1982).

Much more significant as a proportion of total system costs are the costs of the service mains (table 4.3). Costs are low for commercial and industrial service mains because of the large average building size. For dwellings, there is clearly a relationship with density, but other factors are important too. The low values for tenements and some terraced housing, for example, are due not only to high density but also arise from the ease of routing mains through internal stairwells in tenements and through roof space (or basements) in some terraces. Flats, on the other hand, present difficulties for internal mains routing. The high cost of service mains for semidetached housing probably vindicates the exclusion of 'the suburbs' from the feasibility studies for 'lead city' schemes.

Table 4.3. Service mains costing coefficients (source: Atkins and Partners, 1982).

	Cost per m^2 of floor space (£)
Terraced housing (buried connections)	17.563
Terraced housing (roof space connections)	9.077
Flat	15.111
Tenements	9.206
Semidetached housing	34.69
Commercial	1.541
Industrial	1.204

All nine schemes were found to be feasible when assessed by discounted cash flow (with a 5% discount rate for the base case). They could provide heat at a cost of 10% below the cheapest alternative whilst showing rates of return of at least the 5% per year required of new investment by nationalised industries, so that they could be comercially viable even if started immediately.

It should by now be clear that although the viability of energy-conserving CHP/DH systems is influenced by spatial structure, and especially by built form, density, and interspersion of land uses, it is not possible to define, for example, specific threshold densities below which such systems would not be economic in the future. At present a figure of about 44 dwellings per hectare has been adopted (in the definition of HDHL areas) but this figure will change with other variables. The relationship between spatial structure and the viability of CHP/DH is dynamic.

Emphasis to date has been on selecting existing areas with the best prospects for viability, and the criteria for investment have been fairly strict—arguably much too strict when the need is to take a long-term view. Indeed, progress in the United Kingdom often seems painfully slow and cautious, especially when compared with what has been achieved in other European countries, like Denmark. But if there is to be an expansion of CHP/DH schemes to meet a higher proportion of heat load in the future, should this possibility be taken into account now in the planning of new areas and in areas of redevelopment or rehabilitation? It is cheaper and simpler to install district heating mains at the time of initial construction than to 'retro-fit' them. But if the viability of CHP/DH in an area seems too far off for this investment to be made, should attention be given now to those aspects of spatial structure which will affect the economics of its introduction in the future? In terms of density, this may not be an issue, since many new developments meet current criteria for CHP/DH under certain assumptions, and the threshold of 44 dwellings per hectare seems more likely to fall than to rise in the future. Local authority housing in the United Kingdom is currently constructed at around 35 dwellings per hectare, and densities of around 30 dwellings per hectare are typical of many new private developments (Turrent et al, 1981, page 61). But layout may be as important as density per se in new areas—for example, it would be much cheaper to connect straight rows of housing than more complicated layouts and, as already noted, costs are greatly reduced when there is scope for internal routing of mains. There is clearly some potential conflict with aesthetic design requirements here. It would also be advantageous to mix land uses, for the reasons already stated, and because a mixture of land uses means that the demand for heat and power is more evenly spread across different types of consumer. In general, the criteria which will facilitate the introduction of CHP/DH at some stage are moderately high densities, built forms which facilitate internal routing, linear layout, and mixing of land uses; but it is difficult to be precise about these criteria and they are certain to change over time.

CHP/DH and the inner city
If attention for the time being at least is focused on existing inner-city areas, then recent trends in population density, housing characteristics, and economic activity cannot be ignored. Some people have suggested that these trends would threaten the viability of CHP/DH even in those areas which presently seem capable of supporting major schemes (see the discussion by Murphy, undated). Others have been attracted by the idea that such schemes may *contribute* to inner-city regeneration. The Chairman of the CHP Group drew attention to this possibility in the foreward of the final report, and the consultants of the feasibility study point out that
> "if CHP/DH proved commercially viable, the availability of relatively cheap and convenient heating, together with the activities associated directly with the development of a CHP/DH network, could make Inner City areas more attractive to both private and public sector investors and influence locational decisions" (Atkins and Partners, 1982, page 50).

CHP/DH might also complement the renewal of the building stock by private and public agencies in such areas.

Pessimistic and optimistic prognoses along these lines are both probably too simplistic. Although housing densities have fallen rapidly, there are still very significant numbers of people living in inner-city areas at densities at which CHP/DH is viable. Redevelopment tends to be of relatively high density, and standards of comfort have been rising. Furthermore, the types of inner-city floor space which tend to be vacant are concentrated in categories such as storage or process industries, which have little or no demand for heat, whereas floor space in tertiary activity requiring heating has probably risen in recent years (even if there has been no net increase in the number of jobs in this sector) (Murphy, undated). The consultants in their summary report suggest that "no net changes in floorspace area by land use will occur during the implementation period of CHP programmes" (Atkins and Partners, 1982, paragraph 5.4). The validity of this statement is obviously based on fairly optimistic assumptions about government policy as well as on judgments based on past trends in inner cities, but if the level of commitment of all the major parties to the inner cities is maintained, current trends do not seriously threaten the prospects for CHP/DH, and it is reasonable to anticipate continued high levels of demand for heat in such areas in future.

CHP/DH might assist inner-city regeneration in three ways. It would create jobs—just under 40 000 man years of employment in the lead city schemes (though some would arise from redistribution of jobs in the energy industries)—and these would have multiplier effects (probably 50–70% of initial employment) in the service industries. It would contribute to improvement in housing quality by providing adequate heat. Last, and less tangibly, a government commitment to CHP/DH schemes in inner cities might stimulate 'confidence', by providing a visible indication that public policy will continue to underwrite the future of these areas (Murphy, undated).

These possibilities require much further analysis. Once again there is a danger in assuming that complex trends might be reversed by relatively simple measures, and although it remains an intriguing possibility that the large-scale introduction of CHP/DH might influence the processes of inner-city decline, suburbanisation, and counter-urbanisation, it seems most unlikely that in isolation it could have any significant and lasting effect.

Solar powered DH?
Finally, it is conceivable that some of the advantages of solar power and district heating could be combined, if solar energy were collected centrally and the hot water distributed through a district heating network. This would have a number of advantages over 'on site' solar heating, including economies of scale, compatibility with higher density housing, and simpler provision of a 'backup' system. Collectors could if necessary be located on the roofs of buildings such as schools and community centres, and on top of the distribution and storage facility itself. It has been estimated that a neighbourhood of 1000 people would need 2600–4270 square metres of collector area, with the distribution and storage facility occupying some 2000 square metres (Smith et al, 1978). Economics of the DH network would be similar to those already considered above. Work in the United Kingdom suggests that large solar heating systems are not at present cost-effective (at least in the conditions prevailing in the United Kingdom), but could become so within the next twenty-five years (MKDC, 1982).

Conclusion
At the local scale the relationship between energy and spatial structure is relatively well defined. We know which built forms are most energy efficient, and how siting, the physical arrangement of buildings, and densities affect the prospects for using ambient energy sources and combined heat and power. Many (though not all) of the structural prerequisites for energy efficiency could be achieved without significant additional cost or loss of amenity.

Perhaps surprisingly, the requirements for different kinds of energy-conserving systems do not turn out to be mutually exclusive. Although CHP/DH (as well as public transport) is favoured by higher densities, whereas the opposite tends to be true of harnessing renewable energy sources, there is considerable overlap, so that densities of, say, 35–45 dwellings per hectare could be compatible with passive (or small-scale active) solar systems as well as permitting the introduction of CHP/DH. The use of *both* systems could be envisaged, but it must be remembered that successful harnessing of solar power on a small scale will reduce the heat load in a given area and could adversely affect the economics of a CHP system. As noted, solar district heating is also a possibility in certain circumstances.

Much broader issues are also raised by the prospects of new energy technologies. In the long term, any commitment to extensive use of 'soft' energy resources would have spatial implications greatly transcending the local level, almost certainly requiring some form of decentralisation. More immediately, if hopes that major infrastructural investment could help to regenerate the inner cities are well founded, then this energy system could indirectly influence important urban trends. These broader issues are surrounded by uncertainty, however, and can really be identified only as interesting research topics.

Meanwhile, what can be concluded is that when all (energy) factors are considered, moderate to high densities emerge favourably, since these encourage energy efficiency in a number of different ways, but need not preclude the use of renewable energy sources. Such densities, combined with mixing of land uses (but local concentration of employment and service opportunities), provide a spatial structure which would be compatible with policies to save energy by promoting public transport, CHP, or small-scale use of renewable energy sources.

Energy-efficient environments: synthesis and policy implications

Introduction
Several approaches to the identification of energy-efficient spatial structures have now been explored. In this chapter the evidence is drawn together, the characteristics of land-use and activity patterns which reduce energy requirements or increase flexibility are summarised, and spatial structures at different scales in which these energy advantages could be achieved are presented. These tend to be 'ideal' structures, assuming a 'clean slate' for development; but if this seems abstract and unrealistic, most of the *principles* at least could be applied to incremental change. Energy savings from modifications to spatial structure are quantified as far as possible, suggesting that potential benefits do merit inclusion of energy considerations in the formulation and evaluation of land-use policies. This leads finally to discussion of the policy implications of what we now know about energy, land use, and the potential for future conservation.

Energy-efficient environments—a synthesis
Efficient characteristics
Although the terms have often been regarded as synonymous, a useful distinction can be made between energy-efficient *characteristics* of spatial structure (in terms of densities, mixing of land uses, etc) and the spatial structures themselves. This distinction raises some interesting questions. For example, could all the efficient characteristics now identified be accommodated within any single spatial structure? Or conversely, can any given set of energy-efficient features be achieved by many *different* spatial structures, or only by one unique form? Energy-efficient characteristics have been identified from exploration of possible responses to fuel constraints, comparison of alternative forms, and definition of the 'spatial requirements' for various energy-conserving technologies. It is apparent that the three approaches lead to a considerable degree of consensus about the attributes of an energy-efficient form.

Work on the possible evolution of spatial structure under energy constraints suggests that reduced energy consumption would result from closer association between different land uses and from higher development densities—features which also emerge very favourably from comparison of alternative forms within an empirical or theoretical framework. This in itself is encouraging because it implies that in promoting energy efficiency in land-use policies, planners would not necessarily be involved in the conflict with market forces which has frequently been predicted (for example, see Pauker 1974). Comparative analysis also suggests that relatively small urban size and linear development patterns have energy advantages, though these are not as well defined as the benefits of greater compactness and closer integration of activities. At the local scale, the case for higher densities is reinforced by the greater energy efficiency of built forms with a low surface area to volume ratio.

The characteristics of an environment which can be thought of as 'inherently' energy efficient, with low useful energy requirements, have much in common with those which allow maximum opportunity for fuel-conserving technologies. Compactness and mixing of land uses (with clustering of trip ends) facilitate trips by public and nonmotorised transport and allow the economic introduction of energy-efficient CHP/DH systems. Perhaps surprisingly, these characteristics are not incompatible with the small-scale exploitation of solar power, though passive solar systems require increasing design ingenuity at densities of more than about 35 dwellings per hectare. However, to meet a high proportion of a community's energy needs from ambient energy sources would require low overall densities (unless renewable sources were exploited using very centralised high technology). Thus, on first examination it seems that some forms of renewable energy future may be precluded by the kind of compact spatial structures which have many other energy advantages. Dependence on renewable sources is normally treated as a remote possibility, especially in the forward planning of the industrialised countries, but given the relative permanence of physical infrastructure, it might still be ill-advised to close the option. The potential conflict between keeping this option open and exploiting the energy advantages of higher densities might, however, be capable of resolution within a 'linear grid' spatial structure described in more detail below.

In summary, the *characteristics* which emerge most consistently from the different approaches to identifying energy-efficient forms are compactness, integration of land uses, and clustering of trip ends, together with at least some degree of 'autonomy' or 'self-containment' at the sub-urban scale (since many of the energy advantages are lost if individuals do not make use of those facilities which are physically close to them). In theory, these features would lead to a reduction in useful energy requirements for travel and space heating, and increased efficiency of primary energy consumption, because of their compatibility with particular transport modes and energy supply technologies.

Efficient structures
Energy-efficient characteristics can, in fact, be exhibited by a range of spatial structures at different scales. The number and size of relatively self-contained urban units is variable, though as they become smaller, the possibilities for 'autonomy' decrease or demand increasing changes in life-styles. Within these units there are various ways in which development can be arranged to achieve the same overall characteristics in the interests of energy efficiency.

At one extreme is the 'compact city' idea where high densities and integration of activities would be achieved by containing the functions of a large city within a very small area of land, usually by means of a multistorey enclosure. This is exemplified by the structure proposed by Dantzig and Saaty (1973), designed to house 200 000 people in a circular weatherproof enclosure eight levels high (figure 5.1), or the essentially similar concept explored by Courtney (1976) as one radical alternative to meeting growth centre needs in Australia. Travel and heating requirements are reduced to a fraction of those associated with more conventional urban forms.

The compact city idea has generally, and not surprisingly, been greeted with scepticism (for example, see Steadman, 1977; Van Til, 1979). Quite apart from the serious questions which could be raised about the flexibility and sociological implications of such a form (especially given the sobering past experience with high-rise high-density development), its apparent energy advantages do not stand up to detailed scrutiny. The totally artificial environment of the city would demand a high-energy input for lighting, cooling, and air conditioning; the opportunities for using ambient energy sources on a small scale would be negligible; and the initial energy 'investment' in the structure would

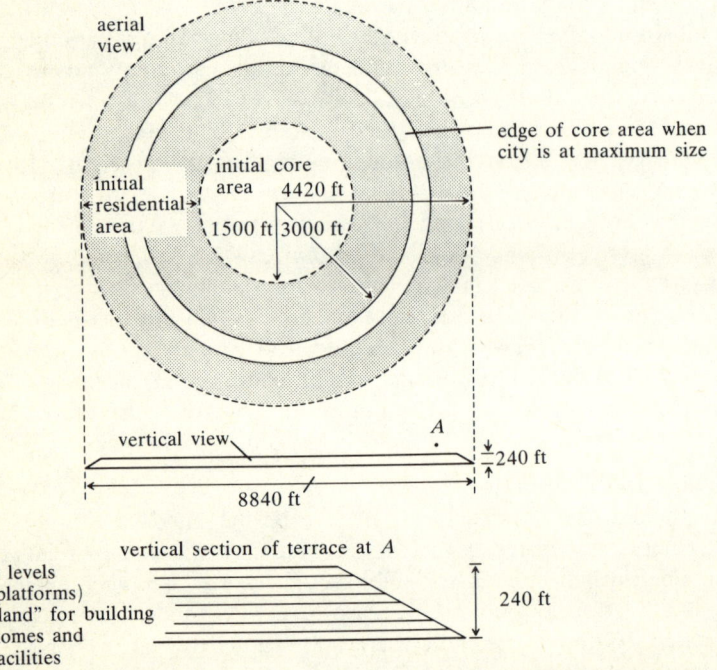

Figure 5.1. 'Compact city' (source: Dantzig and Saaty, 1973, © W H Freeman).

almost certainly be very high. Its inhabitants might also use large quantities of fuel to escape as far from it as possible during their leisure time! However, in a less extreme form, the centralised high-density city emerges as a plausible energy-efficient option (for example, see Clark, 1974; Edwards and Schofer, 1975; Fels and Munson, 1975; Roberts, 1975). In this form, employment and services are centralised and physical separation is minimised by high-density residential development around the centre. The potential to use renewable energy sources is likely to be limited.

At the other extreme is what Mathieu (1978) has called an 'archipelago' pattern, consisting of compact nucleated urban subunits having 'walking distance' or bicycle scales (figure 5.2). Clearly the population of these areas must be large enough to provide the necessary threshold for a range of facilities if they are to be sufficiently 'autonomous' to reduce overall travel requirements. But it is difficult to define a minimum size since thresholds change over time. In the past they have increased as travel costs have diminished and greater economies of scale have been sought; under energy constraints thresholds may fall again, making smaller units 'viable' and potentially energy efficient. 'Pedestrian scale' clusters of between 10 000 and 30 000 people, up to several hundred hectares, have generally been envisaged (Mathieu, 1978; Thomas and Potter, 1977). These units would not necessarily have to be contiguous, nor indeed need they be conceived of as subunits of an *urban* area, though the more widely separated they are, the more energy will be

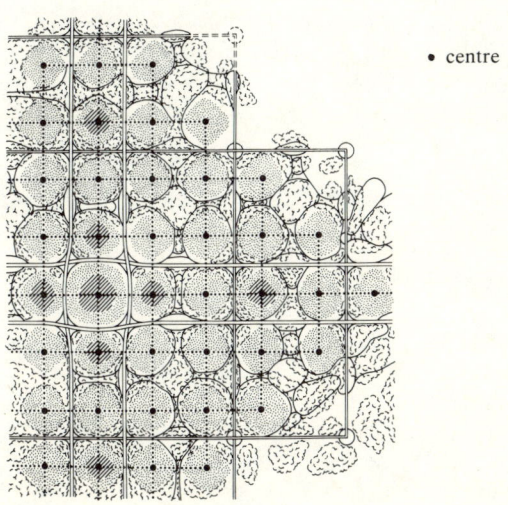

Figure 5.2. 'Archipelago' pattern (nucleated urban subunits) (source: Magnan and Mathieu, 1975).

consumed for any necessary travel between them (figure 5.3). It is worth reiterating the point that this nucleated 'archipelago' pattern may not be energy *efficient* if mobility is relatively unconstrained and people choose not to take advantage of their 'local' employment and service opportunities. But it remains a *flexible* pattern in the sense that these facilities would be available if mobility were to *become* restricted at some stage in the future. In spite of these qualifications, decentralisation of employment and service opportunities has often been found to be more efficient in terms of travel requirements than has concentration in a single centre (for example, Hemmens, 1967; Schneider and Beck, 1973; Stone, 1973), and existing urban areas might be expected to 'coalesce' into a nucleated pattern as energy constraints increase (Albert and Banton, 1978; Romanos, 1978). Comparison of incremental development patterns has also sometimes indicated that growth above a certain level would be better directed to 'secondary' centres than to a single main nucleus (Clark, 1974; Owens, 1981).

It is not difficult to recognise in the concept of 'decentralised concentration' some aspects of the familiar 'neighbourhood' idea much advocated by urban planners in the 1960s, and an integral feature of British New Towns policy. The similarity is not surprising since the neighbourhood concept arose partly from a desire to reduce unnecessary travel and provide 'everyday' facilities at local level. The 'energy-efficient' units could achieve this, but would almost certainly involve more compact development, more decentralisation of employment and services, and greater autonomy of function.

One potentially serious objection to compact nucleated spatial structures is that there would be insufficient access to large open areas to satisfy aesthetic or recreation requirements. Access both to urban and to rural amenities is a long-cherished ideal, particularly in British planning, first encapsulated by Howard in his 'garden cities' design at

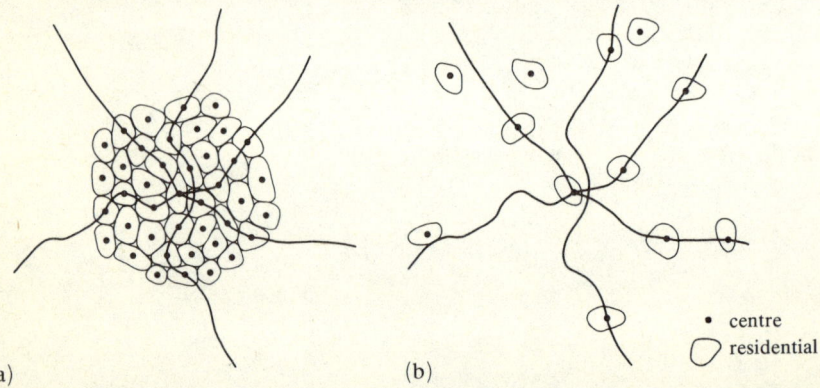

Figure 5.3. Urban subunits: (a) contiguous, (b) noncontiguous.

the end of the nineteenth century (Howard, 1898). It may be argued that the very compactness of the proposed spatial structure would make it simple to escape an urban area into the countryside beyond, as Dantzig and Saaty (1973) claim for their compact city, or that open areas could lie between subunits, which would make the structure similar to Howard's concept of a 'city federation'. Separation of units in this way would, however, incur an energy cost. If more immediate access to green areas is required—perhaps, thinking ahead to a more 'self-sufficient' future, for greater use of ambient energy sources or more urban gardening—then some rethinking of structure at the intraurban scale is necessary if energy efficiency is to be maintained.

The energy advantages of relatively high development densities and integration of diverse activities could be compatible with access to open land and a wider range of life-styles and energy systems, in a linear grid structure, based on the ideas of March (March, 1967; Martin and March, 1972). This permits a high linear density of development[5] in which integration of land uses is achieved by concentrating origins and destinations of trips onto a small number of routes. It is an ideal structure for public transport and CHP/DH systems, but because it is 'full of holes' it would be compatible with even quite extensive use of renewable energy sources (Rickaby, 1979; Steadman, 1977). Figure 5.4

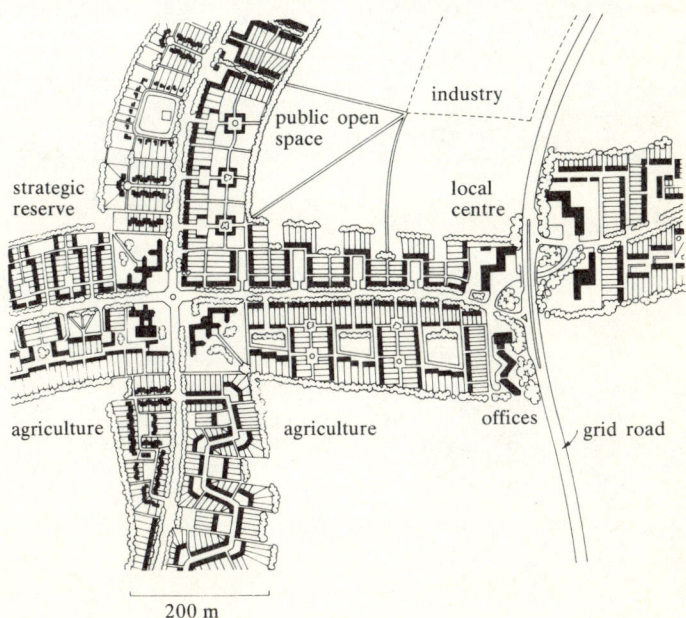

Figure 5.4. Notional 'linear cruciform' pattern (source: Rickaby, 1979).

[5] The concept of linear development is actually independent of density.

illustrates linear development—in this case with a 'cruciform' pattern—and also shows that this structure is compatible with the concept of semiautonomous units at the intraurban scale.

These proposals inevitably raise the spectre of 'ribbon development', the control of which has required relentless planning effort. And perhaps people want to live in 'places', not on lines or grids, so that energy efficiency, low overall densities, *and* the elusive sense of 'community' cannot all be achieved simultaneously. But it would be a mistake to confuse the linear grid with unplanned ribbon development.

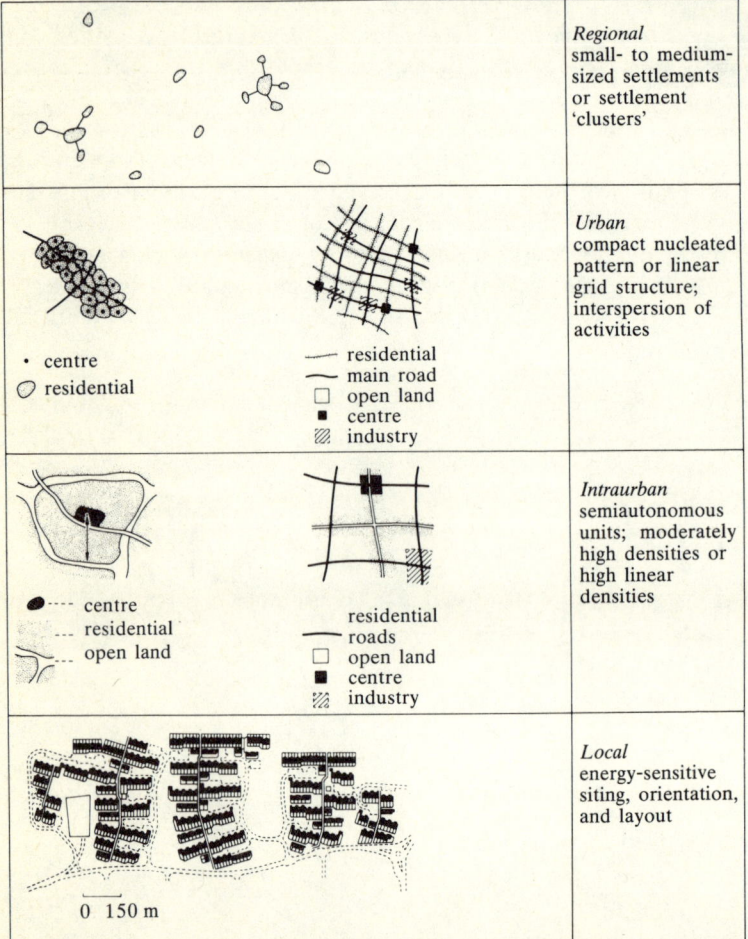

Figure 5.5. Energy-efficient spatial structures at different scales [source: after Owens (1984b); *urban* and *intraurban* after Rickaby (1979); and *local* after a Basildon project, see Turrent et al (1981)].

The 'band' of development could be wide enough to allow dwellings to be located well away from the main route and emphasis could still be placed on local centres "accessible by diverse means from diverse places" (Rickaby, 1979, page 16). This is illustrated by Rickaby's proposals for a new area of Milton Keynes, based on linear cruciform development (Rickaby, 1979; see chapter 6, figure 6.4). Nevertheless, all ideas for energy-efficient spatial structures must quite rightly be subjected to critical scrutiny, and their nonenergy implications will be significant constraints on the potential for their achievement in practice.

In figure 5.5 some of the spatial structures which are potentially energy efficient are summarised. Although discussion of ideal structures tends to assume green field development, many of the principles could also be applied to growth and incremental change. For example, in large cities it would be possible to encourage nucleated development around existing suburban centres and/or along transport corridors, and to ensure that growth resulted in greater integration of activities rather than in greater separation. At the local scale, the criteria for densities, built form, and orientation can normally be applied both to green field and to incremental development, though small in-fill sites may offer less flexibility in this respect.

The potential for energy conservation

Spatial structures have now been identified which are, in theory, relatively efficient in energy terms, but this in itself is not very useful without a closer examination of their various costs and benefits. Before looking at the policy implications, we need some idea of the magnitude of the potential energy advantages, to set alongside the other implications of energy-efficient forms, and to compare with energy savings which could be achieved by more conventional 'nonspatial' means.

The magnitude of savings

Variation in useful energy requirements between different spatial structures is, in theory, quite large, but it must be remembered that many of these results are based on urban models which *assume* a certain behavioural response to energy constraints. The magnitude depends in part on the number of structural characteristics which are allowed to vary. Shape alone accounts for a variation of up to about 20% in travel requirements (Jamieson et al, 1967; Stone, 1973), whereas interspersion of land uses causes travel or transport energy requirements to vary by a maximum of about 130% (Hemmens, 1967; Owens, 1981; Schneider and Beck, 1973). In combination, factors such as shape, density, and the spatial arrangement of activities lead to variations of about 150% in travel or transport energy requirements (Edwards and Schofer, 1975; Stone, 1973). At the smaller scale, significant theoretical differences in energy requirements for space heating between different built forms can easily be demonstrated, with forms like intermediate flats being up to three times more efficient than equivalent detached dwellings (for example, see BRE, 1975).

To the extent that certain spatial structures may enhance opportunities for efficient transport and/or energy supply technologies, these structures will be more flexible, and potentially more energy efficient than others, though it is less clear that energy savings could then be directly attributed to spatial structure in the same way as could reduction of the useful energy requirements, described above. It has been estimated that transfer of 50% of urban work trips and 50% of other personal and social travel from private to public transport (bus) would result in a lowering of energy requirements for these journeys by 8–12% and 16–19%, respectively (Maltby et al, 1978). CHP could improve the efficiency of primary energy conversion in power stations by some 100%, representing significant improvements in the efficiency of energy use for space heating. Careful attention to siting and orientation could ensure that at least 20% of the energy requirements of a typical dwelling were met by 'free' ambient sources (Leach et al, 1979).

The energy implications of different structural variables are summarised in table 5.1. The spatial structures shown in figure 5.5 gain their energy advantages from a combination of the mechanisms given in table 5.1, but it cannot be assumed that the resulting savings are simply additive. Bearing this in mind, the figures suggest that transport energy requirements might vary by a factor of two to three, and space-heating

Table 5.1. Energy implications of structural variables.

Structural variable	Mechanism	Energy implications
Shape	travel requirements	variation of up to about 20%
Interspersion of activities	travel requirements (especially trip length)	variation of up to 130%
Combination of structural variables (shape, size, land-use mix, etc)	travel requirements (trip length and frequency)	variation of up to 150%
Density/built form	surface area: volume ratio affects energy requirements for space heating	200% variation between different built forms
Density/clustering of trip ends	facilitates running of public transport system	energy savings of up to 20%
Density/mixing of land uses	facilitates introduction of energy-efficient CHP/DH systems	efficiency of primary energy use improved by up to 100%
Density/siting/orientation and landscaping	maximises potential to use 'free' ambient energy sources	can reduce conventional energy consumption by at least 20%

energy requirements by a factor of two, between the least and the most efficient spatial structures. Larger differences in transport energy requirements—up to tenfold variations—between different urban forms have been claimed, but are largely accounted for by variation in 'nonspatial' factors such as modal split and efficiency of modes (for example, see Edwards and Schofer, 1975; Fels and Munson, 1975). Taking the first and smaller set of figures (between twofold and threefold variation) implies that a society with a very inefficient spatial structure, which used 16% of its primary energy for transport and 50% for space heating, could expect to reduce its overall primary energy consumption by about one third if it were possible to change to energy-efficient land-use and activity patterns. This is a very significant reduction, but represents an upper limit in the sense that it derives from total transformation of the least-efficient spatial structure to the optimum form. In practice, change is likely to be incremental, the starting point may not be the 'worst case', people may not respond as the models assume, and the optimum structure is unlikely ever to be achieved.

Comparison with other conservation measures
Of the options which have more conventionally been contemplated for reducing transport energy requirements, improvements in vehicle efficiency have generally been considered to be the most promising. A 20-25% reduction in fuel use has been considered feasible by the early twenty-first century (Banister, 1981; Maltby et al, 1978), and Leach et al (1979) think that a 50% reduction should be possible in the United Kingdom by 2025. Beside these figures, the twofold to threefold differences between alternative spatial structures seem not insignificant, especially if it is accepted that with appropriate policies, substantial changes to spatial structure could be effected within a similar time period. Technical and structural changes will both be subject to constraints, though those affecting the latter are generally assumed to be more intractable.

'Nonspatial' measures to reduce energy consumption in new buildings, primarily improved insulation standards, have a large conservation potential. Building regulations assumed by Leach et al (1979) to be enforced by 1990 would result in typical space-heating requirements being halved. Again, the potential savings from higher densities, efficient built forms, and careful siting and orientation compare well, but (with the exception of the last) may be subject to more difficult constraints.

Of course, the effects of different *kinds* of energy conservation measures will not be cumulative. If fuel consumption in cars was cut by half, potential absolute energy savings from reducing the physical separation of activities would be smaller. If building standards reduce space-heating requirements, less can be gained by promoting energy-efficient built forms. This is not an argument to ignore potential measures in any field, but it emphasises the importance of understanding the costs and benefits of adopting different measures. It has often been assumed that modification to spatial structure would have a relatively high cost to benefit ratio, but this has never been demonstrated convincingly. On the contrary, energy-efficient environments seem to offer many other social and environmental advantages.

There are obviously uncertainties about the theoretical potential to reduce energy consumption by modification to spatial structure, even before considering the constraints which will prevent the full potential from being realised. Physical, socioeconomic, and political constraints are immediately apparent and have already convinced some planners that they can have "no key role ... in energy conservation" (Wright, 1979). But the magnitude of potential savings, however qualified, suggests that the possibilities deserve further exploration; and the fact remains that the spatial structure of society and the energy system are closely interrelated. Since planners influence spatial structure it is difficult to accept, as we enter an era in which energy is widely expected to become increasingly scarce and expensive, that they should continue to ignore the energy implications of their plans and policies. It can be acknowledged at the outset that 'ideal' structures are probably unattainable, but potential constraints should not be allowed to preclude at least *consideration* of energy factors in the planning process. The energy–land-use relationship has important policy implications.

Energy and planning: the policy implications
The broad policy implication of the energy–land use relationship is simply stated—energy efficiency should be a legitimate concern of land-use planning at all scales. More specific implications must depend on particular circumstances, especially the available resources of infrastructure, finance, and manpower, and the institutional and legal framework within which policies are formulated and applied.

Authorities ultimately responsible for land-use planning must first accept that it is both possible and desirable to develop energy-flexible land-use patterns and that planning intervention has an important role to play in their evolution. Unless those at the 'sharp end' of forward planning and development control operate within a broadly favourable policy framework, individual initiatives and innovations are unlikely to succeed. Some of the most successful attempts to integrate energy and planning considerations in US cities (as in Portland, Oregon, for example) have been carried out within a broader 'energy-conscious' State policy framework. In contrast, in the United Kingdom, local planning authorities are discouraged from including energy-related objectives and policies in structure plans by an ambiguous and sometimes negative attitude in central government (Owens, 1986).

The extent to which energy considerations are integrated more generally into the various fields of responsibility of local government also helps to define the policy implications for land-use planning. A corporate approach has some advantages over ad hoc recognition of the issue by individual departments. In the USA, 'local energy management programmes' are being developed in a growing number of communities, involving local government, the utilities, developers, industry, and the public. Typically, these involve appraisal of the community's current and projected energy demands, alternative possibilities for supply (including local and renewable sources), a package of conservation measures (*some* of which may involve land use and transport), programmes of public education, and technical and financial assistance. The whole is coordinated by a management or leadership group, including representatives of the different interests (Meshenberg et al, 1982). Within this framework, land-use planning is simply one of a whole series of measures aimed at achieving the objectives of the energy management plan.

In the United Kingdom, many local authorities have appointed 'energy managers', but the concept of the role is narrow, usually restricted to energy efficiency within local authority buildings such as offices and schools. Control over energy supply and over the various fields of local authority administration is more centralised than in the USA, where energy is supplied by private utilities and there is greater local autonomy in relation to such functions as, for example, building regulations. In the United Kingdom there is less scope for integration of relevant policies at local level to form an 'energy management programme' or something similar. In attempting to implement energy-related policies without such a framework, planners are bound to feel frustrated by their limited powers. They can exercise little control, for example, over siting and orientation of buildings. These considerations are usually ignored because they are no one's particular responsibility, yet potential energy savings are large and costs are small. There is a case for extending planning control to such matters or for ensuring that they are taken account of in the building regulations.

Within an appropriate policy framework, energy considerations should form an integral part of forward land-use planning, and relevant policies would be implemented largely through the normal means of selective public investment and development control. For the sake of illustration, energy considerations at each stage of a familiar model of the planning process [Hall (1974), adapted from McLoughlin (1969)] are shown in figure 5.6 (see over). (The fact that planning is not really like this model in practice is one of the many constraints on effective integration!)

Details will depend on the way in which investment decisions are made and on the means of regulation of public and private development, but broad guiding principles are apparent from the energy-efficient characteristics already identified. The kind of matrix shown in table 5.2 (see over) would also be useful at the policy formulation stage. Generally speaking, the aim should be for compactness and integration of land uses, within semiautonomous units which may be small in number and relatively large, or numerous and relatively small. Where a large growth increment is planned, there will be more choice over the number, size, location, and contiguity of 'subunits'; otherwise existing centres may provide nuclei for further growth. A broad range of policy options is open even when energy efficiency is an explicit objective, but in any case relative energy efficiency should be among the criteria for evaluating alternative development patterns. Some particularly important considerations for this stage are listed in table 5.3.

Accurate quantification of energy implications might be viewed as a problem, but it should not normally be necessary. The analogous experience with environmental impact assessment has demonstrated that overenthusiastic quantification of impacts, even where possible, is not a useful input to decisionmaking which must ultimately depend on professional and political judgment (Brooks, 1976; Clifford, 1979).

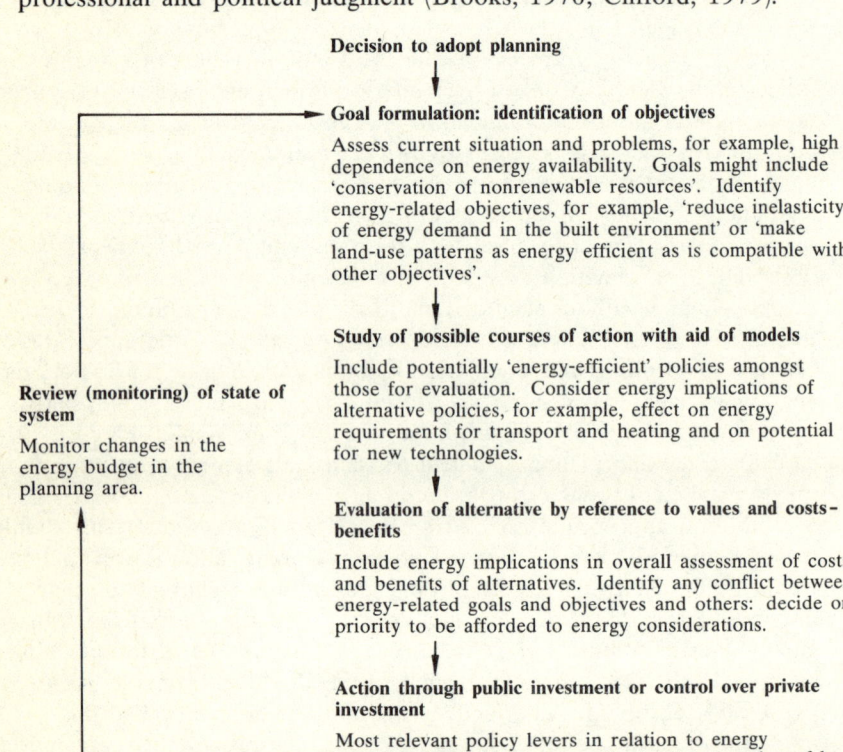

Figure 5.6. Planning process with energy considerations.

Table 5.2. Simple interaction matrix to show how energy–spatial structure relationships might be identified[a].

	A	B	C	D	E	F	G
Energy requirements for transport	×	×	×	×			
Energy requirements for space heating		●	●	×	×	×	
Potential for efficient public transport system	×	×	×	×			
Potential for walking/cycling	×	×	×	×			
Potential for introduction of CHP/DH	×	×	●	×	×		
Potential for use of renewable energy sources				×	×	×	
Energy 'investment' in infrastructure	×	×	×	×			
Energy requirements for industry		●					×

A location and degree of dispersal of residential development
B location and degree of dispersal of industrial development
C location and degree of dispersal of services
D density of development
E built form, layout, orientation
F siting in relation to microclimate
G industrial structure
[a] × interaction, ● possible interaction.

Table 5.3. Sample checklist of energy considerations for use in the study of alternative planning policies.

1. How will the policy influence travel requirements—for journeys to work, to shops, to schools, to recreation, etc? Will the physical separation of activities be reduced or increased? Will people be trapped in expensive travel patterns if fuel prices rise? Does policy encourage or discourage the use of the private car, public transport, walking/cycling?
2. Does the location of new development take sufficient account of microclimate?
3. Would the development densities implied by this policy
    (a) encourage or discourage the use of energy-efficient built forms,
    (b) permit the use of passive solar energy for space and water heating,
    (c) permit the economic introduction of DH or CHP/DH now or in future?
4. Is there any potential, apart from control over densities, to influence built form, orientation, and layout of buildings to maximise energy efficiency at the 'microlevel'?
5. Does the policy encourage decentralisation of employment and services? If so, how does the loss of 'economies of scale' compare with potential savings in transport energy requirements? How likely are people to use local facilities?
6. Is new industry likely to be energy intensive? Could less energy-intensive industries be encouraged by industrial location, or any other policies?
7. How do the energy implications of this policy vary in the short/medium/long term (as far as it is possible to judge)?
8. Could the policy be modified in any way to improve energy efficiency whilst still meeting all the other objectives?
9. In comparison with alternative policies under consideration, is this policy energy efficient/energy intensive/robust?

Some idea of the relative magnitude of energy requirements in policy alternatives, together with an indication of their 'flexibility' in relation to energy supply and demand, should show whether they have significantly different energy implications and would be a great improvement on the present situation, which is often one of total ignorance. A related consideration is that planning authorities already have access to much of the information which would be needed in considering the energy implications of alternative land-use patterns, in the form of survey and other data gathered in the normal course of the strategic and local planning process.

Finally, some of the problems of fitting a new type of assessment into everyday planning practice, and acquiring the relevant expertise, could be overcome by taking the issue out of the normal planning sphere in the initial stages. For example, Cheshire County Council in the United Kingdom has a 'long-range futures' group to consider longer term issues which might be 'squeezed out' by everyday local authority activities. One of the issues which has been examined by this group is the role of the planning department in relation to the energy system in the County (R Lowe, personal communication, 1981).

Constraints
The deceptive simplicity of the process outlined above brings us once more to the question of constraints. Planning practice bears little resemblance to any neat model, and new procedures must be integrated into an already complex and inherently 'messy' process. The problems of physical and institutional inertia, resource and information constraints, and the limited effectiveness (or quite unintended results) of planning policies must already be abundantly clear, and will be further illustrated in the context of practical experience. The other potential difficulties (again, not limited to energy considerations) involve possible social, economic, and political conflicts. With the exception of the latter, which probably represents the most significant single constraint, these problems may well have been exaggerated. In the remainder of this chapter, it is suggested why.

These and other constraints are best illustrated in the context of practical experience, reviewed in the following chapter. But this is a good point to comment on three of the most obvious problems to be faced by energy-conscious planners—inertia in the built environment, conflict with other objectives, and conflict with vested interests.

Economic constraints
One question which must arise is whether energy-efficient planning policies will prove 'expensive' in more conventional economic terms. This is inherently unlikely since on the available evidence, development patterns with energy-efficient characteristics such as 'compactness' tend

to be those which can be serviced most economically and in which travel costs can be kept at a relatively low level. At the smaller scale, it has been shown that many simple spatial adjustments (orientation, for example) can be made to achieve significant energy savings at little or no extra cost.

But there is some conflict, at least in the short term, on the issue of mixing of land uses. Some degree of decentralisation of employment and services is needed to minimise energy requirements. The authors of 'Europe 2000', a futuristic look at a more resource-conscious continent, argue, for example, that

> "in most circumstances, small units, well distributed among the population are preferable to large units requiring long journeys to reach them. Offices, shops and factories should all be planned so as to be readily reachable by foot or bicycle; the era of the giant factory, the huge office, even the great urban complex is perhaps over" (Hall, 1977, page 247).

But recent trends have been towards greater centralisation in order to gain advantages from 'economies of scale'. Cost savings, choice, and enhanced opportunities, at least for some groups in society, have tended to weigh more heavily than additional costs, most obviously travel (and therefore energy) costs, but also social costs, particularly for less mobile members of the community. The new problem in this calculation is the uncertainty of future energy prices—indeed, this is the rationale for attempting to incorporate energy considerations into the planning process in the first place.

Making some simple assumptions, it can be readily demonstrated that the more travel costs are assumed to rise, and the lower the rate at which future costs and benefits are discounted, the greater the relative benefits of decentralisation of facilities (table 5.4, see over). A similar dilemma arises when energy-related objectives involve capital expenditure now to achieve energy savings in future; this is one of the main obstacles to the introduction of combined heat and power systems in the United Kingdom. Assumptions about the discount rate and future energy costs are crucial—the lower the discount rate and the higher the projected increase in real fuel prices, the more options which involve expenditure now to save energy in the future will be favoured.

In view of the considerable uncertainty about the future price and availability of energy resources, the normal discounting procedure which attaches little importance to energy costs in twenty or twenty-five years time is arguably inappropriate; there is a 'risk' element to energy-intensive options which is not fully incorporated into their costs. This is especially the case when new infrastructure will last well beyond the date when costs are reduced to insignificance in such an analysis! In this context, it is worth reiterating the theoretical finding that certain land-use patterns are 'robust', in that they have relatively low energy requirements over a range of assumptions about future energy costs. There is a strong case for planning policies to promote these patterns, as long as they are acceptable in other ways.

Table 5.4. Costs and benefits of centralisation.

Assumption[a]	Case A: one large shop	Case B: three small shops
Benefits (assuming loss of economies of scale in case B)	100 units per annum (notional units)	75 units per annum
Initial travel costs	20 units per annum	negligible
1 Travel costs constant, 10% discount rate		
NPV benefits	936	702
NPV costs	187	0
benefits − costs	749	702
2 Travel costs increase at 10% per annum, 10% discount rate		
NPV benefits	936	702
NPV costs	400	0
benefits − costs	536	702
3 Travel costs increase at 10% per annum, 5% discount rate		
NPV benefits	1308	981
NPV costs	644	0
benefits − costs	664	981

[a] NPV is estimated for a twenty-year period using the formula

$$\text{NPV} = \sum_{t=0}^{19} \frac{Vt}{(1+i)^t},$$

where t is the year, i is the discount rate, V is the cost or benefit in year t.

Social and environmental considerations

Energy-efficient development would certainly be inhibited by social and environmental constraints if it could be achieved only by the arrest or reversal of powerful forces which are presently shaping society. Current trends, it has often been argued, reveal a preference for low-density living and certainly for living away from the larger urban areas and away from industrial development of any kind (for example, see Edwards and Schofer, 1975; Keyes, 1976; Pauker, 1974). Thus, any move towards higher densities would be resisted on amenity grounds, and the concept of closer integration of different land uses tends to raise a spectre of 'dark satanic mills'. In the view of one author, sceptical about the move towards greater energy-efficiency, "the very idea of closeness to work is probably not compatible with the nature of advanced industrial and post-industrial societies" (Pauker, 1974, page 15). Modern industries like electronics are not 'bad neighbours' in the traditional sense, but the values and prejudices of many decades may prevent the adoption of policies aimed at integration of land uses, even where this could be achieved with little or no sacrifice of environmental quality.

On closer examination, the potential conflict between current trends and those deemed desirable for a more efficient use of energy resources turns out to be something of a paper tiger. Current trends originate in a period of cheap fuel, and are not immutable. Theoretical work suggests that if energy costs rise, forces will be set in motion which would tend in any case to move the built environment towards a more energy-efficient structure. Although energy constraints will not *necessarily* result in spatial adjustments, especially in the shorter term, planners are likely to be working with underlying forces rather than against them. Nevertheless, planning should anticipate problems before they are generally apparent, and it is still possible that policies will be opposed if they are perceived to be counter to individual preferences. It must be demonstrated, therefore, that energy-efficient planning policies do not conflict with other social and environmental goals.

The emphasis on resistance to increasing densities in the literature on energy and land use seems to arise because so much of what is written refers to the USA, where typical densities are very much lower than those in the United Kingdom and where urban sprawl has been encouraged by the relative absence of controls on land use. The single-family home on a large suburban plot is very much an integral part of the 'American dream'. For Americans, therefore, development at higher densities will require significant changes in attitudes and trends, but in the United Kingdom most new development is already at densities quite compatible with energy-related objectives, and this particular constraint is unlikely to be a problem.

The related fear that an energy-efficient development pattern would be one with little amenity or aesthetic appeal has almost certainly received too much emphasis. This is largely because of the erroneous but common assumption that energy efficiency implies high-rise high-density development, possibly even along the lines of the nightmarish 'compact city'. More considered evaluation of alternatives reveals that this certainly need not be the case. There exists a variety of spatial structures in which energy efficiency can be compatible with a range of other social objectives. At the sub-urban and local scales, nucleated or linear grid structures could provide accessibility and amenity as well as flexibility for a range of different energy futures. At the regional scale, an energy-efficient pattern seems to be one of many small- to moderate-sized towns. Counterurbanisation in developed economies suggests that this settlement pattern must also be attractive in other ways.

In the United Kingdom, although a small number of county planning authorities include explicitly energy-related policies in their structure plans, the majority have been 'unconsciously energy conscious' by including objectives, policies, and evaluation criteria which implicitly promote energy efficiency, but whose main purpose is social or environmental (Owens, 1986). For example, integration of land uses, one of the main criteria identified for a potentially energy-efficient environment, is an objective of many plans, but primarily for nonenergy reasons. In Cornwall, minimising the time and cost spent in travel is seen as particularly important in a county with one of the lowest average earning levels in the country; in South Yorkshire the aim is to provide greater opportunities for the employment of those unable to travel far to work. As would be expected, policies to achieve greater mixing of land uses vary with circumstances. In many urbanised counties, such as Bedfordshire, Nottinghamshire, and Tyne and Wear, new development will be concentrated in existing urban areas, and the Greater Manchester Plan goes further in seeking to encourage high densities, "as far as they are compatible with amenity", and in regarding industry as compatible with residential areas, "as long as it is without serious environmental effects" (GMC, 1981, policies H5 and EC12). Some planning authorities in more rural areas aim for integration on a smaller scale. For example, the planning authority in Gloucestershire, in a reversal of previous policy, will encourage dispersal of services to smaller communities to help stem out-commuting (GCC, 1981). Again this demonstrates the variety of ways in which energy-related policies (whether explicit or not) may be pursued.

Broadly speaking, the evidence is that development patterns which are desirable from an energy viewpoint are also being sought in the United Kingdom for their social, economic, and environmental advantages. If these different objectives are pulling roughly in the same direction, then with a little more attention to detail in relation to the energy issue, policies to achieve an energy-efficient and otherwise attractive environment can be formulated. Conflict between energy-related and other social objectives does not then emerge as an insurmountable constraint to the integration of energy considerations into planning practice.

Political constraints
The most significant constraints are likely to arise not because energy efficiency is incompatible with other goals, but because energy-related *and other* policies run counter to vested interests and will encounter strong political opposition. Planning policies attempting to guide growth to selected areas have often been resisted by developers on the grounds

that they are too restrictive, or by adverse public opinion in 'growth' areas. In Norfolk, for example, villages around Norwich, identified for growth in the draft County Structure Plan, vigorously defended themselves against further development, whereas villages in remoter rural areas protested strongly at the presumption against further growth (NCC, 1977). In many areas, development lobbies are exerting strong pressures for relaxation of greenbelt policy, though this is undesirable on energy and environmental grounds (for example, see Fleming and Short, 1984; CPRE, 1984).

There are also strong pressures to build 'out-of-town' car-orientated facilities for shopping and entertainment, running counter to the concept of more decentralised local facilities which meet the objectives of energy efficiency and accessibility for the less mobile. This exemplifies the more general problem that although planners can refuse some applications and make it known that they will encourage others, they cannot *make* developers come forward with applications which they do not perceive to be advantageous. Planning authorities often complain about 'lack of positive planning controls' (Owens, 1986). Neither private developments like retailing nor public services are directly under the control of planners, and considerations which are difficult to quantify, such as possible future energy constraints, will continue to be outweighed in the planning of facilities like shops, hospitals, and schools by the more readily defined 'benefits' of economies of scale. In areas with high unemployment, more or less *any* development promising jobs is welcomed, whether or not it is compatible with energy-related (or other) planning policies. Even in a growth county like Berkshire, in one of the fastest growing areas of Britain, where new industries often seek green field sites inaccessible except by private car, planners are reluctant to resist such development lest they should prejudice opportunities for enhancement of employment in the growing industries (R S Stoddart, personal communication, 21 December 1983). The intensely controversial nature of many transport policies has already been noted, and provides a further example of the impotence of planners confronted with powerful political lobbies.

All of these pressures are difficult to resist during plan formulation, especially when groups involved are influential and articulate in the public participation phase. During plan implementation too, policies are difficult to uphold when there is little support from central government for the philosophy behind them. Ultimately, county planning authorities in the United Kingdom do not have the power to implement their strategic planning policies, since local authority development control decisions may be overturned on appeal to the Secretary of State for the Environment.

Concluding comments
Having identified a set of spatial structures which are potentially energy efficient, it is not difficult to recognise the planning policy implications both at a broad general level and in particular local circumstances. There are quite clearly many constraints, physical, social, and institutional, which make it difficult to incorporate new ideas into the planning process and prejudice the implementation and effectiveness of policies. These can seem so overwhelming when listed that the prospects for energy-integrated planning might seem hopeless.

But in spite of the difficulties, many planning authorities are beginning to take the issue seriously. A number of cities in the USA have now adopted or implemented energy-efficient land-use policies. In Scandinavia, considerable progress has been made in the integration of energy and land-use planning. In the United Kingdom, although there have been some interesting initiatives, there has not been such a marked commitment. About a quarter of the county structure plans include explicit energy-related objectives, policies, or evaluation criteria, but these tend, on the whole, to provide additional justification for policies which would have been pursued in any case. It is encouraging, however, that the response to a survey of local planning authorities on this issue was very positive, and many replies indicated that greater attention would be given to energy considerations during reviews of structure plans (Owens, 1986). The main obstacles seem to be lack of a clear view from central government and lack of information, neither of which should prove insurmountable.

In conclusion, it can be said that the constraints to energy-integrated planning are not unique. They can be overcome at least as much as can other constraints familiar to land-use planners, given the willingness to proceed and provided that it is accepted that many imperfections will remain and that startling results are unlikely to manifest themselves in the short or even the medium term. Many planning authorities have taken the first important step of recognising the significance of the energy issue, and some have begun to formulate and implement relevant practical policies. It is these first attempts to include energy considerations in the planning process that are considered in the final chapter.

Energy-integrated planning in practice

Introduction
A considerable amount of academic research on energy, spatial structure, and planning has been carried out within the last decade. Although understanding remains imperfect, there is an encouraging degree of consensus about the characteristics of energy-efficient forms, and the policy implications of this theoretical work are beginning to emerge quite clearly. Many planners have recognised and have expressed interest in the potential contribution of land-use planning to energy conservation. In some areas, as yet regrettably few in number, attempts have been made to translate theory into policy and practice, and positive steps have been taken to integrate energy considerations into the land-use planning process. Examples in this chapter are drawn from the United States of America, the United Kingdom, Denmark, and Australia. Davis, California, and Portland, Oregon, are among the best known examples of energy-integrated planning. In the United Kingdom there has been little progress in this direction, but some county planning authorities have produced 'energy-conscious' structure plans and, on a smaller scale, one of the remaining areas of the new city of Milton Keynes is to be planned as an energy-efficient community. In Denmark, land-use planning and planning for energy supply have become closely integrated; Århus County provides a useful illustration of the issues and constraints involved. The final example is drawn from Australia, where quite radical energy-related planning policies are being advocated for Melbourne and other communities, though as yet they remain to be accepted and implemented.

Where it has been adopted, energy-conscious land-use planning has often been an integral part of a much more comprehensive programme of energy-conservation measures; this is the kind of 'package' which has been implemented in a number of cities in the USA. The approach has much to recommend it, but it does create difficulties for evaluation since any effects of 'spatial' measures cannot be considered in isolation. But in any case, since modification of spatial structure can normally be expected to reduce energy consumption only in the medium to long term, it is still much too early to assess the effectiveness of recently implemented energy-conscious planning policies.

Before discussing actual examples of energy-integrated planning, it is interesting to consider why some communities have perceived its potential benefits and have actively pursued them, whereas in other areas the energy issue is hardly recognised, or is afforded very low priority. To some extent differences may be explained by the prevailing conditions of energy use and efficiency in different countries. It is hardly surprising, for example, that more visible progress should have been made in the USA, with the highest ratio of energy consumption to gross national product in the world, than in the United Kingdom, where land-use patterns are already much less energy-intensive and the scope for improvement commensurably smaller, though still significant.

Another explanation for lack of progress in the United Kingdom may be the generous endowment of energy resources, and the large number of energy policy options, in comparison with a poorly endowed country like Denmark, where there has been a greater imperative for conservation. But national factors are not the only explanation; within the USA, for example, there is great variation in the level of commitment to energy conservation in land-use planning, and in the United Kingdom, structure plans display widely different levels of 'energy consciousness' in objectives and policies.

Lee (1980) has examined the factors which induced seven American cities to establish comprehensive energy-efficiency programmes, including in most cases relevant land-use planning policies. (The cities are Davis, Los Angeles, and San Diego in California; Seattle, Washington; Portland, Oregon; Wichita, Kansas; and St Paul, Minnesota.) He identifies four essential factors.

"First there is a strong leadership by senior elected officials on behalf of the program, especially Mayors, who in each case are willing to assume a degree of political risk in altering the traditional agenda. Second, there is a relatively greater tradition of open government and local resource management. Third, there is a pervasive belief that improving the energy efficiency of the building stock, and of the city in general, will be in the long run economic interest of the city. In other words, each city believed that a program to promote energy efficiency would complement their economic and urban development goals. Fourth, each city established an extensive participatory process which served to legitimise the issue in the eyes of the electorate" (Lee, 1980, pages 26–27).

Another factor was also important for five of the cities, which had faced some kind of crisis providing extra impetus for their energy programmes. Seattle and San Diego had rejected plans for new nuclear power capacity and needed an alternative policy; Los Angeles decided to act after the 1973 oil embargo; the economy of Wichita suffered severe disruption in the 1977 winter gas shortages; and Portland had been subject to shortages of both natural gas and hydroelectricity, the latter due to a drought in 1972–1973. These problems helped convince officials and the public that improving energy efficiency would reduce their vulnerability to disruption and improve the quality of life. As Lee (1980, page 30) notes, "This attitude constrasts with the pervasive notion elsewhere that such programmes would cause a deterioration in life style and unnecessary hardships".

Many cities have some kind of energy programme, but most have not been prepared to allocate substantial resources to it, because of lack of leadership, traditional perceptions of the role of local government, and unwillingness to take political risks or to reallocate resources from other areas.

In the United Kingdom most planners who have decided to give energy considerations some attention cite uncertainty about future energy supply and prices (emphasised by the energy crises of the 1970s) as justification for doing so—reducing vulnerability is the most prominent consideration. That some planning authorities have considered these issues while others have not probably reflects, as Lee found for the USA, the attitudes of particular individuals in the planning process and their perception of the relative importance of energy issues and the extent to which they are linked to other social problems. In the case of the Milton Keynes project there is also quite clearly a more positive desire for demonstration and publicity behind the proposals.

One factor which has undoubtedly deterred consideration of energy issues in UK planning departments is a generally unhelpful policy framework. In 1976, the Department of Energy rejected a recommendation from the Select Committee on Science and Technology that the energy implications of land-use planning policies should be clearly identified (Department of Energy, 1976). Structure planning regulations and guidelines (for example, Department of the Environment, 1974; 1981) do not encourage integration of energy considerations, and the Secretary of State for the Environment has deleted energy-related policies from some plans when they have come before him for approval, on the grounds that they are 'nonland-use' policies outside the terms of reference for strategic planning. Typical examples of deletion include a policy in the submitted plan for Clwyd, seeking to locate development so as to minimise travel and conserve fuel (CCC, 1980, policy number 65), and a policy in the submitted Greater Manchester Plan, which aimed to maintain a public transport network capable of short-term expansion "should this be required in the event of fuel shortages" (GMC, 1979, page 80). Not surprisingly many authorities have interpreted the legislation and the attitutde of Central Government to mean that energy is not a legitimate strategic planning issue, in spite of its fundamental relationship to the spatial structure of society. As the County Planning Officer for Leicestershire explained:

"We have found to date that the scope for developing specifically energy-related policies is somewhat limited, given the rather broad brush nature of the plans and in particular the rather negative advice from the Department of the Environment" (D R Sabey, personal communication, 31 January 1984).

Although the Departments of Energy and Environment have acknowledged informally (in response to a query from the Greater Manchester Council) that energy conservation is "a major structural consideration" (GMC, 1983), the situation remains confusing, and planning authorities in the United Kingdom lack advice and encouragement on this issue.

The cases considered below further illustrate the factors leading to the formulation, adoption, and implementation of energy-conscious planning policies; they also illustrate many of the constraints discussed in the previous chapter and show how, in some cases, these have been overcome.

The USA: Davis, California, and Portland, Oregon
Davis

Davis is one of the best known examples of a city which has taken energy 'on board' in its physical planning. In the 1973 General Plan for the City (City of Davis, 1973), a goal was set for reducing energy consumption by the development of multimodal transport, new building and siting regulations, a more compact city, and public education. This plan, according to the Community Development Director for Davis, "changed the shape of the city from sprawling suburbia to a well-managed, compact community" (McGregor, 1977, page 16).

Davis had sound foundations for an energy-efficiency programme. The need to control the growth of the city was accepted by the City Council in 1968, after a lengthy debate, long before energy became an important issue; the growth management system in the General Plan was not, therefore, alien to the existing planning philosophy in the city. Similarly, in the mid-1960s, provision of bicycle lanes became an important political issue, and Davis had been developing an extensive bicycle lane network since that time (Ridgeway and Projansky, undated).

In 1973, the City Council sought federal grants to produce
1. an energy building code,
2. a building code workbook,
3. a 'retrofit guide' for homeowners,
4. an energy education guide,
5. passive solar designs,
6. a guide to energy-efficient community planning,

the last of which has since formed the basis of the energy policy for the city.

The energy-conservation ordinance was adopted in 1976. Much of the emphasis is at the building scale, where design standards now require a high degree of thermal insulation. The city has sponsored development of 'passive solar' houses, and in areas developed under the General Plan, about 75% of the plots have a north–south axis and houses must be a minimum distance from fences and other structures on the south side. A development in which solar buildings have been combined with an environment designed to enhance amenity and the quality of life is reputed to have been a 'resounding success' (Craig, 1982). Streets in Davis must be as narrow as possible while serving their necessary function, saving on energy 'investment' in infrastructure and reducing traffic speed and fuel consumption.

On a larger scale, the growth management system in the General Plan aims to ensure that development considered 'necessary' for the community is provided in the locations, and at the rate, which can be most effectively serviced by the city, thus controlling urban sprawl and producing a more compact community. It is claimed that "Savings here are obvious—in capital costs, maintenance costs, and energy costs" (McGregor, 1977, page 17), but it is not clear whether these 'savings' benefit the nation or only the inhabitants of Davis. In as far as the management system ensures the rational *planning* of growth, there are clearly genuine reductions in the use of resources; but *displacement* of growth may simply mean that the expenditure is made elsewhere. Particular emphasis in the Davis Plan is, however, placed on provision of low-income housing, which "removes ... the sting and flavour of exclusion which might otherwise be present" [McGregor, quoted in Ridgeway and Projansky (undated, page 41)].

Encouragement for nonmotorised transport also plays a major role in the energy programme for Davis. The unofficial logo of the city is an old-fashioned bicycle and there are more than forty miles of cycle ways and street lanes for cyclists. One in every four miles travelled is estimated to be by this mode (Williams, undated). It should be pointed out, however, that facilities in Davis are nearly all within a fifteen minute bicycle ride, and the climate and the flat terrain make cycling particularly attractive. In different conditions it could be much more difficult to persuade people to use this mode of transport, even if facilities are provided.

Although it is very early to judge the effectiveness of the energy-conservation programme in Davis, preliminary figures from the local utility, the Pacific Gas and Electricity Company, suggest that per household use of electricity decreased by 22% between 1974 and 1982, and per household use of natural gas decreased by 43% (W Williams, personal communication, 12 September 1983). Figure 6.1 (see over) shows the reduction in gas and electricity consumption in Davis in comparison with two similar communities, Woodland and Vacaville (City of Davis, 1982). Energy use has declined more in Davis, and the difference is very marked in the case of electricity consumption. [Other communities were subject to mandatory state thermal standards in buildings from 1977; Davis subsequently amended its own building code to make it even more stringent (Lee, 1984).] It is very difficult to attribute the reduction in energy use to particular causes, though one study suggests that residential electricity demand in Davis over the period 1976-1979 was 15% less than would have been expected in the absence of the energy building codes [Dietz and Vine (1982), quoted in Lee (1984)].

It is even more problematic to attribute any of the savings to spatial modifications, and there are no figures available to show any decrease in transport energy requirements.

Craig (1982, page 518) argues that, "it remains something of a mystery why one city should decide to do so much while hundreds of similar towns are doing so little". But in fact the willingness to innovate in Davis relates well to the kinds of factors identified by Lee (1980). Lee describes the senior elected officials in Davis as "visionary" and willing to take political risks, and suggests that

"It is no accident that a disproportionate number of developed effective energy programs are located in the west. These cities are newer and their traditional perception of the role of local government has evolved from a more recent period in our nation's history" (page 28).

Davis is a small university town whose population (30 000) is likely to be liberal and progressive, and receptive to the philosophies of intervention and conservation. Emphasis in the 1973 General Plan is on conservation of land, water, energy, and other natural resources, moving towards a general limit on growth; energy consciousness fits comfortably into the tradition of local resource management, characteristic of many Western cities in the USA. As Craig (1982, page 519) himself points out, "Energy consciousness played an important role in the complex decisions to move the town in these unusual directions, but concern over values was even more important".

Involvement of the public in debate about the energy ordinances has also played a significant role. The building code was initially opposed strongly by developers who considered that it would lead to expensive, aesthetically unpleasing housing, and that orientation requirements would mean lower densities and reduced profits. Their fears proved unfounded, since housing costs increased by only one or two percent

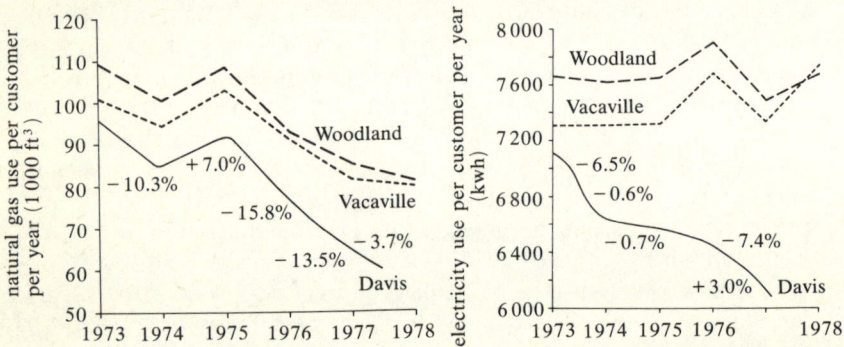

Figure 6.1. Comparative energy use per capita—Davis, Woodland, and Vacaville (source: Lee, 1984).

and there was no change in appearance. But an important effect of resistance to the new building code was to stimulate 'fiery' debate and to focus a great deal of public attention on energy (Williams, undated). The result was to legitimise the aim of energy conservation and to make many of the new 'restrictions' acceptable.

Energy conservation in Davis is already being hailed as a success story (for example, BBC Radio 4, 1984; Lee, 1980; McGregor, 1977). The experiment of the city has shown that energy-conscious land-use planning policies can be formulated and accepted, though it is too early yet to say how effective they are. The final word on Davis should go to the Environmental and Energy Officer for the city:

"it is important to note that the programs which have worked for Davis were created as a response to a set of climatic, political and economic conditions. Davis' most important contribution has been to prove that energy conservation does not have to mean self-denial" (Williams, undated).

Portland

A second example of a city which has been prepared to devote substantial effort and resources to energy conservation is Portland, Oregon. The city and the four surrounding counties which make up the Standard Metropolitan Statistical Area (SMSA) have a total population of about 1 075 000, expected to rise to 1 600 000 by the year 2000. In some ways, therefore, the challenge of improving its energy efficiency is greater than that presented in Davis. In further contrast to Davis, where the initiative came largely from within, the adoption of energy-related policies by Portland took place within the framework of federal support and active encouragement from the state of Oregon, one of whose mandatory statewide planning goals is that

"Land and uses developed on the land shall be managed and controlled so as to maximize the conservation of all forms of energy, based upon sound economic principles" (Oregon Land Conservation and Development Commission, 1975).

Again the efforts of individual elected officials were crucial, and the city has a tradition of open government and close communication between elected officials and the public (Lee, 1980). Largely through the efforts of its mayor, Portland was chosen as a model to demonstrate how a community could identify and assess alternative strategies for energy conservation, as part of a project funded by the US Department of Housing and Urban Development. The crises faced by the city in 1972–1973 added an extra impetus to the energy initiative.

In the Portland Energy Conservation Demonstration Project, completed in 1977, eighty-five possible strategies were examined and forty-two were identified as being applicable to Portland (City of Portland, 1977; Hemphill, 1977; Sewell and Foster, 1980). It was estimated that energy demand in the Portland region could be reduced by 35% in 1995 (compared with projected demand in the 'status quo' option).

One of the chief objectives of the Project was to examine the links between energy saving and urban form. Table 6.1 shows the list of ways to save energy and ways to implement the savings, identified in the

Table 6.1. Ways to save energy and ways to implement savings identified in the context of land use and transport in Portland (source: City of Portland, 1977 volume 3b).

Ways to save
Twenty-seven specific energy-conservation programmes were designed for Portland. Each was examined in this section with respect to the following objectives.
Locate land uses so that people can travel shorter distances for each trip.
Locate land uses so that more people can ride public transit and bicycles or walk rather than use their automobiles.
Promote policies which given preferential treatment to public transit, car pools, bicyclists, and pedestrians.
Promote policies which deemphasize private automobiles use.
Promote policies within the city government which increase the gas mileage of the municipal fleet.
Encourage the construction of housing along major transit streets, near major employment areas, and near shopping centres.
Encourage construction of more high-quality small homes, condominiums, and apartments.
Encourage 'in-fill' housing on substandard and vacant land within the city.
Encourage the conversion of large single-family houses to include a rental unit.
Locate new retail business and office buildings in areas that are well served by public transit.
Prevent the expansion of strip development.
Encourage the location of neighbourhood grocery stores in residential areas.
Limit the height and bulk of commercial buildings.
Discourage energy-intensive or energy-inefficient industries from locating within the city.
Encourage the development of joint sites where secondary industries, office buildings, or housing can use waste heat from other industries.
Locate new industrial jobs and houses near one another.
Centralise truck terminals, contain the tariff and free delivery zones, and encourage the shipment of goods by rail rather than by truck.

Ways to implement
In this section three means of implementing the various ways to save energy are discussed.
Educational programmes aimed at convincing people that they will benefit from conserving energy and telling them how to do it.
Incentive programmes designed to encourage energy savings either through money savings or through governmental zoning, budgeting, and administrative actions which make it easier for citizens to ride public transit or for developers to use land more efficiently.
Mandatory programmes designed to legislate energy efficiency through the establishment of requirements and standards.

volume concerned with transportation and land use (City of Portland, 1977, volume 3b). A simplified procedure for mapping areas of the city according to energy efficiency was also established, resulting in an 'energy zone map' which divides the city into five zones based on relative energy efficiency. The object was to guide new development to energy-efficient locations (table 6.2). For example, it was estimated that if most new development occurred in zone 2 rather than zone 4, Portland would use 7% less energy for transport and 1% less residential energy.

Table 6.2. Criteria for establishing the energy zones of Portland (source: Sewell and Foster, 1980).

Energy zone	Zone characteristics	Efficiency ranking
Zone 1: areas served by transit, shopping, and jobs	sewered areas in Portland that are *either* within 0.125 miles of 1995 transit routes or 1.3 miles of 1995 transit stations *and* within 0.333 miles of major shopping centres *and* within 0.5 miles of industrially or commercially zoned land; *or* within 1 mile of the central business district	excellent
Zone 2: areas served by transit and shopping, or transit and jobs	sewered areas in Portland that are *either* within 0.125 miles of 1995 transit routes or 0.333 miles of 1995 transit stations *and* within 0.333 miles of major shopping centres; *or* within 0.125 miles of 1995 transit routes or 0.333 miles of 1995 transit stations *and* within 0.5 miles of industrially or commercially zoned land	good
Zone 3: areas served by transit, or shopping, or jobs	sewered areas in Portland that are *either* within 0.125 miles of 1995 transit routes or 0.333 miles of 1995 transit stations; *or* within 0.333 miles of major shopping centres; *or* within 0.5 miles of industrially or commercially zoned land	fair
Zone 4: other sewered areas in Portland	other sewered areas in the city of Portland that are not in zones 1, 2, or 3	poor
Zone 5: unsewered areas and areas outside city boundaries	unsewered areas in Portland and all areas outside present city boundaries	very poor

In accordance with this project, and with extensive public involvement, Portland developed an energy conservation policy (City of Portland, 1979), which was adopted by the City Council in 1979. It included commitments to develop land-use policies which would use density and location to reduce the need to travel, and to improve the efficiency of the transport system and reduce its consumption of nonrenewable fuels. These policy commitments were accompanied by more specific objectives relating to the location of new developments and encouragement of energy-efficient transport modes.

A serious attempt was made to incorporate these strategies into the comprehensive plan for the city (City of Portland, 1982). One of the central features of the proposed plan was a 'centres and corridors' concept, which would involve concentrating most new development over twenty years into centres of existing commercial activity with some higher density housing, and along major streets ('corridors') with high-density housing. There would be emphasis on public transport and bicycles. Between the corridors, town houses and, further back, single-family dwellings would be allowed (though still at relatively high densities). Thus, the emphasis was very much on energy conservation through compact high-density development, using an urban pattern which combines nucleated centres with high linear densities. Energy savings were anticipated because in theory less travel would be required, necessary travel would be carried out by more energy-efficient modes (public transport and bicycle), and higher density development would permit more efficient use of fuel for space heating (Sewell and Foster, 1980).

The plan was examined in public during the late 1970s. It won both strong support and, since it represented major change for many of the residents of the city, considerable opposition (Sewell and Foster, 1980). Its energy-related features were among the most controversial. The division of views is illustrated in table 6.3, showing greatest support from environmentalists, newspapers, and heads of local government agencies, and opposition mainly from developers, industrial and union leaders, and the Chamber of Commerce. Zoning, which can drastically alter land values, was particularly unpopular with development interests and landowners. The emphasis on high-density corridors was strongly opposed by some neighbourhood groups, with the result that the planmakers retreated by breaking up several corridors into clusters of commercial activity, surrounded by high-density housing, which were termed 'development nodes'. This well illustrates the point that any plan which tries to take energy considerations into account will be subject to intensive lobbying by interest groups, the power of which must be recognised as a major constraint on energy-related objectives. It is interesting to note that the division of groups in favour and opposed to the Portland plan bears some resemblance to that found in recent research on modern environmentalism. Cotgrove and Duff

(1980; 1981), for example, show that groups on what they term the 'periphery' of corporate capitalism, such as academics, planners, and other public employees, are more likely to hold environmental values than are those at the 'centre', including central government, the major corporations, and the trade unions. This division of views is to some extent reflected in attitudes towards planning in general and to energy-conscious planning in particular.

As with Davis, it is difficult to tell how effective these policies will be in the longer term in modifying spatial structure and in reducing energy requirements. The Portland experience demonstrates that it is possible to plan for greater energy-efficiency in large established urban centres; but political opposition may result in considerable 'watering down' of such plans. Of greatest significance, perhaps, is the demonstration value of cities like Portland, which show that progress can be made towards energy conservation and encourage other cities to take similar steps—President Carter sent letters to 600 US mayors pointing out the merits of Portland's energy policy (Sewell and Foster, 1980). The Urban Consortium, a coalition of thirty-seven major urban governments, has developed a programme which is attempting to consolidate the advantages of both of the Davis and of the Portland approaches (Lee, 1984).

Table 6.3. Portland area special interest groups listed in decreasing order of their support for energy-efficient land use (source: Sewell and Foster, 1980).

1	Environmental groups
2	Newspapers
3	Democratic Party
4	Heads of local government agencies
5	City and County employees
6	Radio and television stations
7	Republican Party
8	Bar Association
8	Labour unions
8	Neighbourhood improvement groups
8	Bankers and executives of financial institutions
8	Church leaders
13	Ethnic groups
14	Industrial leaders
15	Retail merchants
16	Chamber of Commerce
17	Other businessmen

The United Kingdom
Strategic planning
In the United Kingdom there has been no equivalent to the experience of US cities which have pursued aggressive energy-conservation programmes. Energy conservation is not perceived as a local issue,

except in the rather limited sense that local authorities are applauded if they reduce their internal energy consumption. But it would be unfair to say that energy considerations have been totally ignored in the land-use planning process. For example, about 20% of the County Structure Plans[6] contain explicitly energy-related objectives, policies, or evaluation criteria. These usually involve reducing travel and, it is assumed, energy use in transport, by integration of different land uses, and concentration of new development. In the Dorset Structure Plan, for example, it is claimed that

"The attempt to provide more jobs and to maintain shops and other services in the rural areas will help to achieve [energy conservation] ... by reducing the need to travel" (Dorset CC, 1983, page 89).

Similarly, in the Devon Plan it is suggested that its policies of providing job opportunities close to home and support for public transport may generally "allow society to make more sparing demands on energy" (Devon CC, 1981, page 149). A handful of plans contain other energy-related policies. For example, in Powys the authority "is anxious to see building design, layout and siting take proper account of climate" (PCC, 1979, page 78). The Greater Manchester and West Midlands County Structure Plans both contain policies to discourage low-density development on energy grounds (GMC, 1981; WMC, 1980).

These, however, are isolated examples, both in terms of planning in general and in terms of particular policies within individual plans. The general picture is one in which energy considerations are not integrated into the land-use planning process; where they *are* explicitly mentioned in plans, they tend to be incidental to the main purpose. Several planning authorities, including Dyfed, East Sussex, Greater Manchester, Greater London, Merseyside, and Powys County Councils, are considering giving energy greater prominence in future, but as yet it remains a side issue in forward land-use planning in the United Kingdom.

In some areas, the small scale and incremental nature of new development is a deterrent to the integration of energy considerations into the land-use planning process. Some planning authorities clearly feel that there is little point in giving much attention to energy considerations when only marginal changes can be effected by their policies, and in some cases the scope is further diminished by the extent of inherited commitments (Owens, 1986).

This attitude may seem justified by the fact that in England and Wales it would take 800–1000 years to renew the existing housing stock at

---

[6] Since the abolition of regional planning councils in 1979, broad planning issues in England and Wales must be considered at the subregional (county) scale. Planning authorities (regional authorities in Scotland) must formulate policies for the development and use of land, focusing on issues of 'key structural importance'. Structure Plans provide the framework for local plans and for development control by district planning authorities.

recent rates of demolition (JLRC, 1984). However, the potential to improve energy efficiency is rather greater than this figure suggests, since the rate of new construction, at approximately 1% of the housing stock per annum, is much greater than the rate of demolition. The bulk of new construction is needed to cater for new household formation, and some is required because of the changing economic structure of the UK economy and consequent geographic shifts in development (table 6.4). This means that in a planning period of about twenty years, new residential development would represent about one fifth of the original stock; but the rate of change will be smaller in the large cities and in the older industrial areas and greater in the growth areas, for example, the 'M4 corridor' and East Anglia. The differential requirements for new infrastructure likely to arise from geographic shifts in industry and population have not yet been quantified. On average, however, the rate of change in the housing stock is small but significant, and there is no good reason to neglect the energy implications of limited development opportunities, as long as it is accepted that large reductions in energy consumption will not be immediately possible.

Table 6.4. Projected turnover of housing stock in England and Wales (source: JLRC, 1984)[a].

	Annual construction requirements	
	1984–1991	1992–2000
To accommodate new households	145 000+	100 000
To replace demolitions, etc	60 000–80 000[b]	40 000–50 000
Geographic shift of industry and other factors		40 000–50 000
Total[c]	220 000	200 000
Total as % of current stock (19.3 million dwellings)	1.14	1.04

[a] The recent construction rate is 178 100 dwellings per annum [mean 1981–1984 (estimated)]. The recent demolition rate is 21 800 dwellings per annum (mean, 1980/81–1982/83).
[b] Figure includes construction requirements in order to replace demolitions and because of geographic shift of industry and other factors.
[c] Approximate values.

Greater London
The activities of the Greater London Council (GLC) deserve special mention. It has probably been the most 'energy conscious' of the UK strategic planning authorities, and the only one to make any serious attempt to document energy flows within the planning area. In the draft revisions to the Greater London Development Plan (originally published in the late 1960s containing no reference to energy conservation), there

is a substantial section devoted to energy supply and demand in Greater London (GLC, 1983). A separate policy statement on energy supply and use in London and on the adoption of a set of objectives to guide the activities of the Council is being prepared at the time of writing.

The Council's objectives for energy policies include the following:
"To promote a comprehensive, secure and efficient system of energy supply having regard to long term opportunities and constraints and to the wider social and economic benefits of alternative energy policies";
and
"To promote the efficient use of energy and the appropriate use of fuels having regard to their intrinsic qualities" (GLC, 1983, page 118).
Policies include a presumption in favour of CHP/DH schemes in high-density heat load areas, justified because "the benefits which would accrue to Greater London as a result of the development of CHP/DH schemes are so considerable that it is appropriate for the plan to provide a planning framework which is favourable to them proceeding" (page 120). There is also explicit support for the development of renewable energy resources and a recognition of the land-use implications. Policy EN9 states that
"In considering the design and layout of new developments, attention will be given to the potential contribution of renewable energy" (GLC, 1983, page 121),
and it is explained that "consideration of the potential role of renewable energy in any new development will be secured through development control procedures" (page 121).

Policies for energy conservation include one which is explicitly related to spatial structure at the 'microlevel':
"Due consideration should be given in the detailed design and layout of all residential, commercial and industrial developments to the desirability of using energy efficiently. In particular, attention should be given to built form, building design, materials, orientation, overshadowing, topography, vegetation, microclimate and other considerations insofar as these affect the consumption of energy" (GLC, 1983, page 122).

There are no specific policies linking energy and spatial structure at the broader scale, but it is recognised that
"the structure and distribution of land uses in London has an effect on the demand for energy, particularly for transport. The precise influence of urban form on energy consumption is still being explored and the situation in London will be monitored and further studies ... undertaken to enable the Council to respond appropriately" (page 123).

The issue has clearly been treated seriously, but constraints on time and resources have curtailed the amount of attention which it will receive.

In the view of a senior GLC planner

"I had very much hoped that we would be able to carry out an evaluation of alternative distributions of homes, jobs, and services during the plan preparation period to see what effect changes would have on energy demand. I did not have the staff resources to do this in the time available and we did not then have the data from the latest transport survey, but I still think it would be very worthwhile. I would start from the existing situation and see what effect a 5 and 10% shift of jobs to central London, to suburban centres, and to home working would have. I suspect that beyond 10% we would have to begin reconstructing the transport system and the whole exercise would become far too complicated" (D Hutchinson, personal communication, 25 January 1984).

The draft alterations to the Greater London Development Plan were the subject of public consultation in early 1984. The energy-related policies were on the whole well received by the London Boroughs, by other organisations including the Ecology Party, the Town and Country Planning Association, and civic societies, and by members of the public (S Gibbs, personal communication, 30 July 1984). However, some Boroughs questioned the relevance of the policies in a plan concerned with land use and transport, or considered that they were premature or too difficult to implement. The policy aimed at 'microlevel' spatial planning (EN13), for example, was supported strongly by some bodies and individuals, was considered inappropriate to the plan by the London Boroughs of Barnet and Havering, and was thought to 'lack teeth' by the Ecology Party (GLC, 1984). Responses also reveal some concern from statutory undertakers that the GLC, in integrating energy considerations into its strategic planning, is seeking to usurp their functions. The Central Electricity Generating Board complained that the GLC listed objectives which were not their responsibility, and seemed unhappy about the emphasis on CHP/DH (GLC, 1984). The alterations are now being revised, but it seems likely that the energy-related policies will survive largely intact, with greater emphasis on their land-use aspects and some modification to accommodate other criticisms (S Gibbs, personal communication, 30 July 1984). Whether the policies will be practical to implement, and to what extent they will be pursued when the planning function of the GLC transfers to the Boroughs after its abolition in 1986, remains to be seen. It would be a great pity to see the effort which has gone into energy-integrated planning at strategic level simply dissipated.

Local energy-efficient planning

On a relatively small scale some isolated attempts at 'energy-efficient' planning have been made in the United Kingdom; these involve the design of housing developments taking microclimatic and passive solar energy considerations into account.

One such scheme at Basildon, for 420 houses on a 10 hectare site, employs a pattern of north–south access roads with short east–west cul-de-sacs (figure 6.2). This enables all houses to be orientated within ±22° of due south. Careful attention has been paid to the spacing between blocks to minimise overshading, and landscaping has been carried out specifically to reduce wind speeds on the face of buildings. Energy considerations have not been allowed to outweigh other important social, technical, aesthetic, and economic factors (Turrent et al, 1981).

A second example using microscale spatial structure to enhance energy efficiency is the Pennylands Housing Project in the new city of

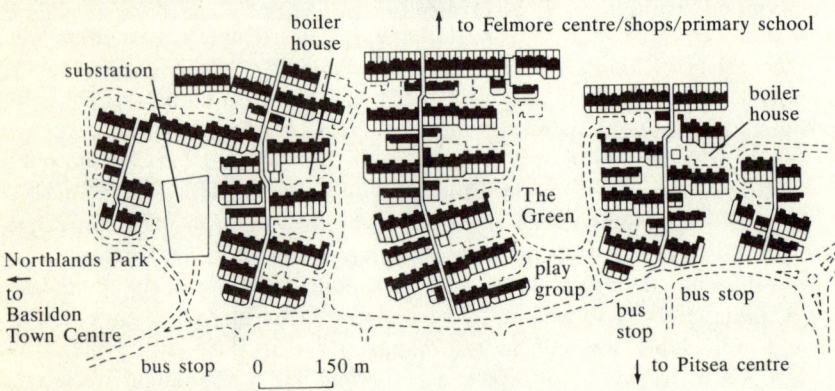

Figure 6.2. Basildon housing site layout. Architects: Ahrends, Burton and Koralek (ABK) (source: Turrent et al, 1981).

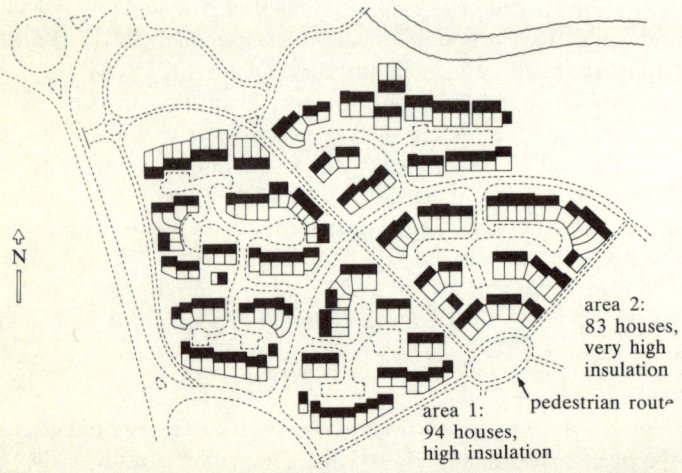

Figure 6.3. Pennyland housing layout, Milton Keynes (source: MKDC, 1982).

Milton Keynes. This consists of 177 houses designed by the Milton Keynes Development Corporation (MKDC), 75% of which face within ±30° of due south, 20% within ±45°, and 5% are aligned east–west (figure 6.3). Energy consumption in Pennylands houses is being monitored to study the effects of different levels of insulation and passive solar energy techniques. Some compromise with environmental criteria was made in this case. One of the original design options for the scheme involved orientating all the houses due south, but this conflicted with the footpath network—established, together with cycleways, to reduce transport energy requirements—and generally created an unacceptably monotonous environment (Turrent et al, 1981). The eventual compromise was to allow some deviation from strict north–south alignment for dwellings, to the extent that other objectives were met with minimum sacrifice in terms of energy efficiency.

Wider recognition of the relatively simple principles involved in small-scale energy-efficient planning, and of the lack of conflict with environmental quality, is still sadly lacking in the United Kingdom. The growing number of schemes like those described above will provide much needed demonstration and information about efficient urban form at this scale, and should pave the way for the more formal inclusion of such considerations into planning or building legislation.

Energy-efficient planning in Milton Keynes

The planning of a new city provides a unique opportunity to integrate energy considerations at an early stage, although in Milton Keynes, designated as a new city in 1967, with a present population of around 110 000 (ultimately to be 200 000), it is only recently that this opportunity has been recognised.

Milton Keynes was planned as a low-density city (average of fifteen houses per hectare), providing for high mobility by means of a rectangular grid of high-speed roads (MKDC, 1970). It is essentially a city planned for high accessibility by private car, with virtually no attention to possible future energy constraints, and typifies the neglect of energy considerations in the land-use planning of the postwar decades. By the late 1970s, when the city was approximately half complete, the possibility of a more energy-conscious approach to planning the remainder began to be taken seriously. The Urban Design Unit of MKDC commissioned a report from the Open University to identify energy-efficient planning policies for the new city (Rickaby, 1979), the proposals in which drew heavily on the linear grid concept. It was suggested that the existing crescent of development be treated as a linear 'corridor', which would require in-filling between existing separate areas and possibly the use of some reserve sites and 'landscape belts'. An attractive public transport service should then be provided along the corridor. A linear grid pattern of development was recommended for the remaining areas (figure 6.4). The latter idea was also considered by the MKDC Urban Design Unit, who showed how it might be compatible in the eastern part of the city with local considerations as well as with theoretical requirements (MKDC, 1979; figure 6.5). In the report, it is argued that

"The policies contained in this strategy are intended to produce a *robust* city. They are intended to shift the emphasis of the city's development towards the provision of a range of lifestyle options. Transport and land use are the main concerns of the strategy, and in each case the middle of the range of options provided must shift towards fuel conservation. There is a role for Milton Keynes Development Corporation in promoting that shift" (Rickaby, 1979, page 23).

These proposals, although they do not constitute an accepted plan for development in Milton Keynes (and have not been implemented as such), provide an almost unique example in the United Kingdom of energy-integrated planning for a real area on a large scale. They do, however, embody a number of assumptions which are not necessarily

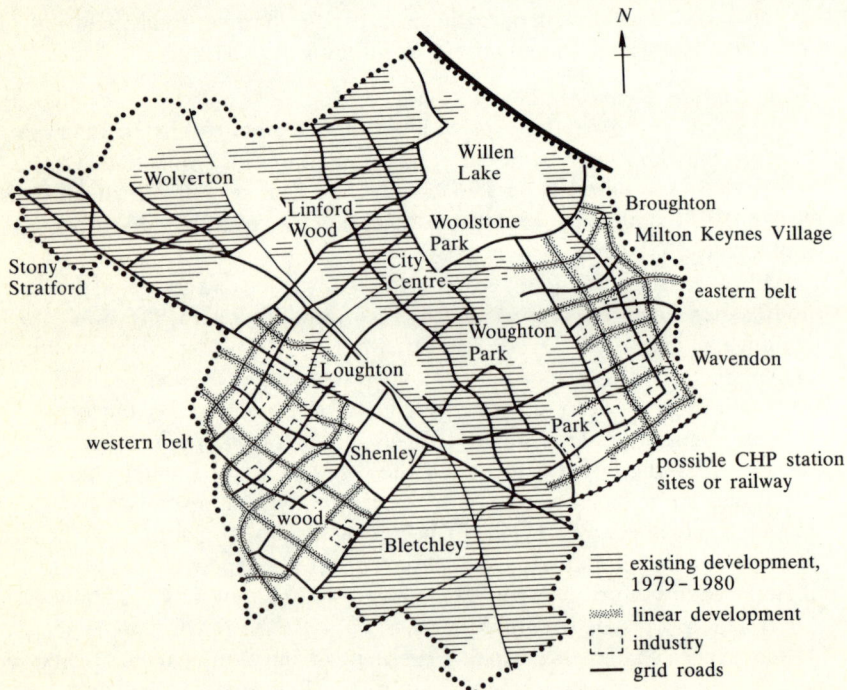

Figure 6.4. Linear belts on the east and west flanks of Milton Keynes (source: Rickaby, 1979).

justified, concerning, for example, individuals' behaviour and the viability of public transport systems. These are considered again below.

Although the comprehensive proposals in the 1979 report were not adopted in their entirety, there now exists a novel plan to design one of the remaining areas of the city as an 'Energy Park'—"an urban area planned from the outset with priority given to policies of energy efficiency" (MKDC, 1983, page 1). The project is being planned and managed by MKDC, who will coordinate the activities of the many organisations involved. It will be financed in a similar manner to other development.

The Energy Park will be an area of some 150 hectares southwest of the City Centre, incorporating a range of land uses including employment (30 hectares), housing (1300 dwellings), open space (30 hectares), and local facilities such as shops and schools. It could be thought of as an urban subunit of the kind which could form the basis for energy-efficient urban structures, but it is surprising that in an area planned for energy conservation, average dwelling densities should be as low as those in the rest of the city (14-15 dwellings per hectare). These densities will permit innovation in the field of renewable energy, but it

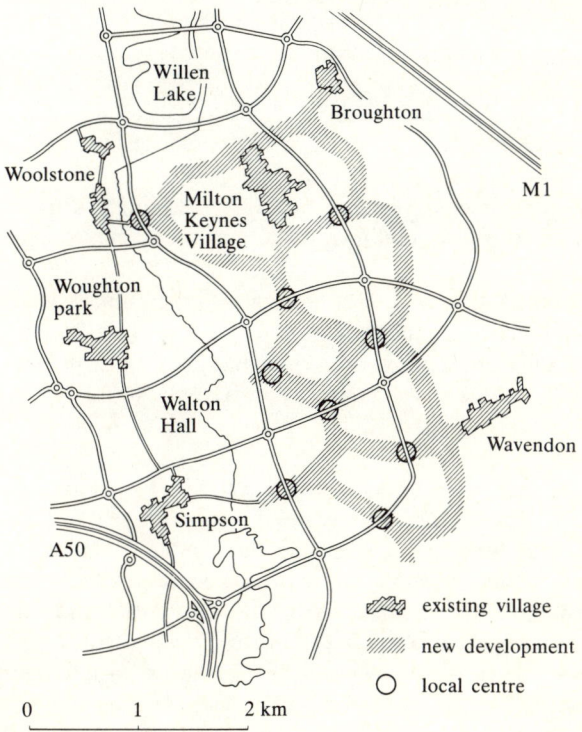

Figure 6.5. A development strategy for the east flank of Milton Keynes (source: Rickaby, 1979).

is acknowledged that CHP/DH is not financially attractive, at least in the short term. It is not yet clear, since detailed plans are still to be drawn up, whether low overall densities will be maintained while achieving some of the benefits of higher densities, by using a linear grid pattern, in the Energy Park.

On a smaller scale, the 'Energy Park' will use the comprehensive approach to energy management typical of some of the American cities discussed earlier [indeed, the inspiration came partly from the energy programme being pursued in St Paul, Minnesota, one of the 'success stories' described by Lee (1980)]. Energy efficiency is to be promoted in a number of ways, including planning and site design, building design, efficient appliances, a range of energy management services, and the provision of information and advice. It is recognised that "Reductions in energy demand can be made through the effective distribution of land use, through building densities and road layouts which enable buildings to be orientated to take advantage of solar gain, and through landscaping" (MKDC, 1983, page 6). There will also be opportunities to demonstrate energy-efficient transport systems. However, a major part of the energy saving is clearly intended to come from energy management, and particularly from the use of innovative information technology. Land-use planning is thus incorporated into a comprehensive energy-conservation programme.

The relatively small scale of the project and its rather isolated nature mean that its main function will be to demonstrate what can (or cannot!) be achieved by energy-conscious planning and management, rather than to make significant energy savings per se. In terms of spatial structure, it will be able to demonstrate possibilities at the sub-urban scale only, with most emphasis on 'microlevel' considerations. MKDC sees the Energy Park as a demonstration project and clearly hopes that it will obtain widespread publicity and recognition. To this end it will include an Exhibition Centre and 'Energy World', a housing area to demonstrate low-energy homes; but the Development Corporation stress that only proven technology will be used—the Energy Park is not intended to develop as a centre for research.

It is difficult to judge the extent to which the Energy Park, construction of which began in 1983 and is expected to take up to ten years, will promote energy-conscious land-use planning in the United Kingdom or elsewhere. Its scale is small, densities are low (in keeping with densities in the rest of the city), and the emphasis is on the local scale and on energy management techniques unrelated to land use. There is also perhaps a danger that the whole project will be viewed as an interesting but unrealistic novelty, which militates against the idea that incorporation of energy considerations should become everyday sound-planning practice at all scales. On the other hand, since British planners have been so slow to recognise their potential role in energy conservation, anything which awakens interest, even in a fairly limited aspect of the energy-land use relationship, is to be welcomed.

Denmark—integration of energy and spatial planning
In Denmark there has been more rapid progress towards energy-integrated planning than in the United Kingdom. In a sense this has arisen out of necessity, since the Danes have little in the way of indigenous energy resources, except for some North Sea oil and gas reserves of uncertain size, and have not yet developed a nuclear power industry. Just before the oil crisis in 1972, virtually all energy needs were met by imported oil (93%) and coal (7%). Faced with an urgent need to reduce oil import dependence, the Danes developed a system of national heat-supply planning with a strong emphasis on energy conservation, formalised in legislation (Act number 258) in June 1979 (Christensen and Jensen-Butler, 1980; IEA, 1981).

Radical changes in the form of home heating are planned and an important element of the scheme is the regionalisation of heating supply by type. Land-use planners have been closely involved in the division of the country into areas where CHP/DH can be introduced (11% of final energy consumption is already supplied by DH), areas which would be better served by natural gas (to be brought ashore from the North Sea in the mid-1980s), and areas where space heating must be provided in some other way. Kommunes (districts) must map present and future energy needs at local level, and the county council draws up a county heating plan which must finally receive ministerial approval. This plan includes the location of specific heat-producing or heat-consuming activities and details of areas in which particular heat-supply methods will receive priority; councils may control the type of heating in new buildings within specified geographical areas (Christensen and Jensen-Butler, 1980; IEA, 1981).

Progress towards integration of energy and spatial planning in Denmark seems impressive in comparison with the procrastination on this issue in the United Kingdom, but it has not been without its problems or its critics, as illustrated by the case of Århus County (figure 6.6, see over) discussed by Christensen and Jensen-Butler (1980; 1982).

About half the homes in the County are presently connected to DH networks (table 6.5, see over). There are many possible alternative combinations of heating systems for the future, including CHP, DH from boilers (with a choice of fuel), individual gas-fired systems, and electric heating. One major uncertainty is the extent to which CHP can be decentralised; schemes are definitely planned for Århus itself (population 250000) and for Randers, but not yet for any other settlements. Another is the extent of the gas network and the timing of its future development. These uncertainties make it difficult to judge whether the Regional Plan is as energy conscious as it is claimed to be.

In many respects the plan is a model of energy efficiency, being essentially based on 'decentralised concentration'. Policies include the strengthening of small- to medium-sized local centres (20000–30000 population), integration of land uses, higher densities, and direction of growth to centres where there are DH networks. But two particularly interesting questions have arisen in relation to future heating systems and spatial structure. One is whether the heating plan for Århus (in which 90% of homes will be connected to a major CHP scheme) will induce greater centralisation in the region, in contravention of the spirit of the Regional Plan and of views expressed during public debate. The other is whether in fact the plan should *aim* for more centralisation, on the grounds that in the long term, reliance might have to be on CHP and electric heating when natural gas is no longer available. Some observers have argued that greater concentration would minimise the

Table 6.5. Methods of domestic heating in Århus County (source: Christensen and Jensen-Butler, 1982; original source: Århus Amtskommune, 1978).

Heating system (central heating)	Percent	
	Denmark	Århus
Oil-fired, individual	52.3	39.5
District heating	27.3	50.9
Solid fuel and other	4.2	1.6
Oven-solid fuel and oil	16.2	8.0
Total	100.0	100.0

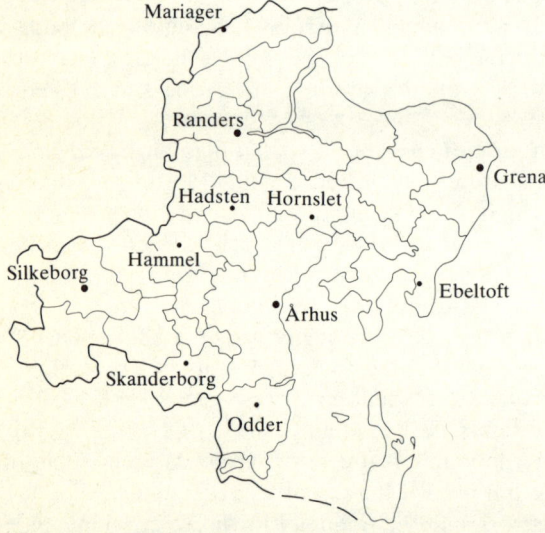

Figure 6.6. Århus County, Denmark (source: adapted from Christensen and Jensen-Butler, 1982; original source: Århus Amtskommune, 1978).

effects of unfavourable developments in the energy field and "ensure the highest possible degree of freedom within the objective constraint of energy availability" (Christensen and Jensen-Butler, 1980, page 47). This is obviously debatable, and there are further uncertainties about the effectiveness of alternative spatial structures in reducing transport energy requirements. In either case—concentration or 'decentralised concentration'—the region will probably be more energy efficient than if no regard had been paid to future energy constraints. The debate illustrates again that there may be no uniquely energy-efficient plan for a given area because of the many uncertainties surrounding future energy prices and availability.

Clearly, the Danes have not solved all the problems of energy-integrated planning, but they have taken some important steps towards recognising the relationships between energy and spatial systems and towards planning simultaneously for their future development. The approach in Denmark is different in emphasis from that in the USA, the former focusing more on the spatial requirements of efficient energy supply systems and the latter aiming for spatial structures which minimise energy requirements for transport and space heating.

Australia—an energy-efficient strategy for Melbourne
A final example of energy-conscious land-use planning is drawn from Australia. It is not an adopted plan, but an idealistic strategy for Melbourne proposed by the Conservation of Urban Energy Group (CUEG) of the Conservation Council of Victoria in the form of a book entitled *Seeds for Change* (CUEG, 1978). The radical proposals in the book are not immediately practical like some plans and policies which have actually been adopted elsewhere, but are important because of the detailed way in which it is shown how an existing city might be made more efficient and because of the attention which is given to changing and, in the view of the CUEG, improving life-styles.

The philosophy underlying the strategy is that urban life-style changes are desirable in their own right and that they can and should be arranged in ways which will be conservative of nonrenewable energy resources. Its proponents argue that it is essential to alter three integrated facets of urban life: transport, land use, and social arrangements. An underlying principle is that of reducing the physical separation of activities—"a process of bringing what-people-do as close as possible to where-people-are" (CUEG, 1978, page 158). This would be achieved through hierarchies of function, density, and transport (figure 6.7, see over). The model presented is one of "supportive neighbourhood houses and lively local foci and district centres strung together with public transport, bicycles and pedestrian access", designed to achieve "a mixed use, people oriented city, developed around public transport modes" (CUEG, 1978, page 311). Other strategic measures would include the gradual increase of urban densities in the built-up areas, improvement to the thermal qualities of buildings, and the reintroduction of open recreational space.

A hierarchy of service centres to meet people's needs sounds very much like that proposed in most planning exercises, but the CUEG strategy involves some significant differences—more offices and light industries at the centres, and higher density accommodation for those who prefer it. This is very close to the concept of 'decentralised concentration', an urban form which has already been identified as energy efficient in theory.

Short-term and longer-term strategies for implementation are presented in the plan, but constraints are given rather scant attention. It is recognised that implementation "will necessarily involve adjustments to the changing maze of taxes, duties, allowances, subsidies, concessions, tariffs, price controls and other regulating machinery that is woven into the economic fabric of our society" (CUEG, 1978, page 377) and "cannot possibly avoid the vicissitudes of the party political arena and will certainly involve extensive legislative and administrative changes" (page 378). But the CUEG have made a conscious decision to avoid

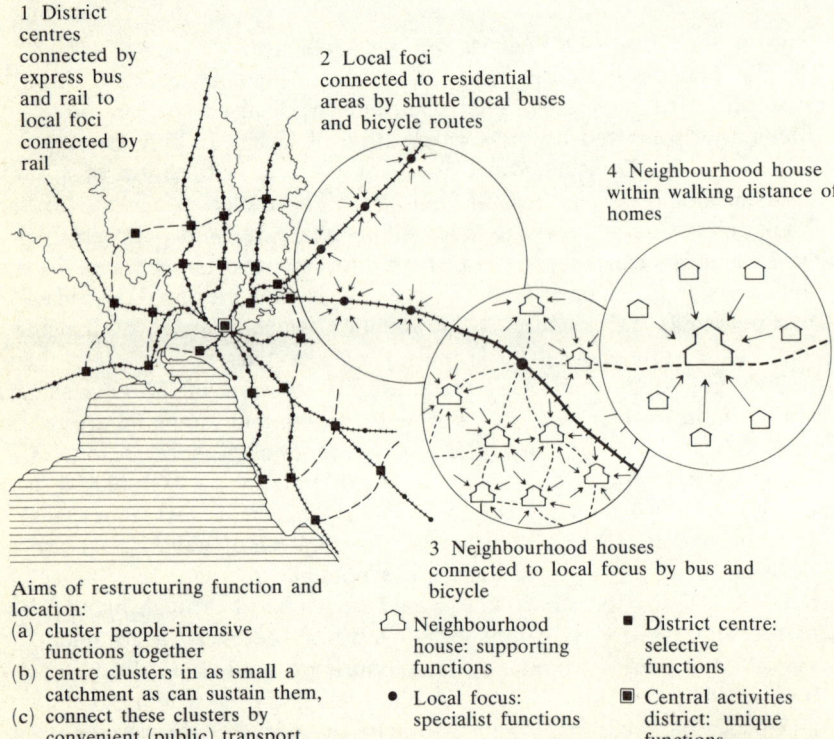

Figure 6.7. Restructuring of Melbourne as envisaged in *Seeds for Change* (source: CUEG, 1978).

"details of this sort" (page 378), because "the all important priority is to determine the reality of our energy problem, a feasible alternative model, and the main strategies that will get us from our present condition to our anticipated model" (page 378). Unfortunately this makes the implicit and highly questionable assumption that 'the energy problem' is separate and distinguishable and can be analysed in an objective way without reference to the policy context in which it is set. It is valuable to postulate what could be done in the absence of policy constraints, since this may itself be part of the slow process of changing long-held views and preferences. But the acid test for 'energy-integrated' planning is whether it can produce practical policies which can be accepted and implemented within the many constraints of the existing system.

Seeds for Change has inspired a number of similar exercises in Australia including a detailed application of energy-efficient planning principles to the City of Nunawading (CUEG, 1979) and to the localities of North and West Melbourne (NMA, 1981). None has yet resulted in definite action by municipal councils, though interest has been shown and the Nunawading Council collaborated in drawing up the proposals for the city.

Conclusion

A review of energy-efficient plans and actual policies is both encouraging and frustrating. It is encouraging that some communities and planning authorities have addressed the issue and are pursuing relevant policies. They remain relatively few in number (though the examples considered here are not the only ones), but this is only to be expected in the initial stages of any innovation. The interesting question—as yet impossible to answer—is whether they will remain isolated examples of energy-conscious communities or whether others will follow suit.

Several factors make a review of current experience frustrating. First, it is too early to evaluate any of the attempts at energy-efficient planning—except perhaps at the very small scale, where the fuel consumption of individual houses can be monitored (as in Pennylands, Milton Keynes)—because significant changes to spatial structure will not occur for a long time and because many other measures are being taken simultaneously to improve energy efficiency. Many of the plans and policies are optimistic about the savings which can be achieved by modifying land-use patterns to reduce travel and heating needs, promote public transport, and facilitate the introduction of new energy technologies. In many cases, ensuring that new physical infrastructure is appropriate may not be a major problem, but it must be recognised that this is only a necessary and not a sufficient condition for reducing energy consumption. In many of the plans, for example, land-use patterns which will 'promote public transport' are sought, because this is potentially more energy efficient than travelling by car. But experience suggests that however attractive a public transport service, the spiral of declining standards and increasing fares may not be reversed unless there are strictly enforced controls on private cars (for example, see Heggie, 1977; Hills, 1983; Starkie, 1982). Such controls have proved very difficult to legislate in most Western democracies, where the 'motor lobby' is a powerful political force (Adams, 1981; Starkie, 1982). Similarly, it cannot be assumed that people will travel less if there is closer integration of land uses. Many factors influencing the energy efficiency of land-use patterns lie largely outside the control of land-use planners. This is not to argue against planning for appropriate land-use patterns, but to advocate caution about *assuming* that energy savings will result. It may be that the effort involved in land-use planning will not produce any measurable energy savings—the frustration lies in our inability to tell at present whether this will be the case.

As Lee (1984) points out, however, in the context of local energy conservation programmes, it would be a mistake to evaluate success in terms of measurable energy savings only. In relation to most spatial modifications this would in any case be impossible in the short to medium term. The American programmes have had an important indirect effect in stimulating energy awareness which itself contributes to a less profligate attitude to energy resources.

More planning authorities in the near future may try to take energy considerations into account. As with all planning policies their success will be limited by many constraints. But the important objective is to move gradually towards an environment in which high energy consumption is not a *necessary* feature of everyday life. Both theory and limited practical experience support the view that there is little to lose by doing this, and there may be a great deal to gain.

References

Anderer J, McDonald A, Nakicenovic N, 1981 *Energy in a Finite World: Paths to a Sustainable Future* Report by the Energy Systems Program Group of the International Institute for Applied Systems Analysis. Program leader: W Häfele (Ballinger, Cambridge, MA)

Adams J, 1981 *Transport Planning: Vision and Practice* (Routledge and Kegan Paul, Henley-on-Thames, Oxon)

Albert J D, Banton H S, 1978, "Urban spatial adjustments resulting from rising energy costs" *Annals of Regional Science* **12** 64-71

Anderson R W, 1973, "Residential energy consumption, single family housing" Hittman Associates, Inc., 9190 Red Branch Road, Columbia, MD 21045

Århus Amtskommune, 1978, "Energi Bilag til rapport om alternative regionplanskiter" Århus Amtskommune, Århus, Denmark

Ashworth G, 1974, "Natural resources and the future shape of Britain" *The Planner* **60** 773-778

Atkins and Partners, 1982, "CHP/DH feasibility programme: stage 1, summary report and recommendations for the Department of Energy" W S Atkins and Partners, Woodcote Grove, Ashley Road, Epsom, Surrey KT12 5BW

Ball D J, Fernandes C, Hutchinson D W, Kostanowicz H, Onslow A, Wright C E, 1981 *Energy Use in London* Statistical Series number 10, Transportation and Development Department and Scientific Branch, Greater London Council, County Hall, London SE1

Banister D J, 1981, "Transport policy and energy: perspectives, options and scope for conservation in the passenger transport sector" DP-36, Bartlett School of Architecture and Planning, University College London, Gordon Street, London WCIH OQB

Banister D J, 1984, "Central-local government relations in Britain: the case of the Fares Fair policy in London" *Transport Policy and Decision Making* **2** 275-289

Barnes D, Rankin L, 1975, "The energy economics of building construction" *Building International* **8** 31-42

Bates J, Roberts M, 1982, "Forecasts for the ownership and use of a car" in *The Future of the Use of the Car* Round Tables 55, 56, and 57, Economic Research Centre, European Conference of Ministers of Transport, 2 Rue André Pascal, 75775 Paris Cedex 16; pp 7-54

BBC Radio 4, 1984, "A town called Davis" 3 March

Beaumont J R, Clarke M, Wilson A G, 1981, "Changing energy parameters and the evolution of urban spatial structure" *Regional Science and Urban Economics* **11** 287-315

Beaumont J R, Keys P, 1982 *Future Cities: Spatial Analysis of Energy Issues* (John Wiley, Chichester, Sussex)

Bellomo S J, Dial R G, Vorhees A M, 1970, *National Co-operative Highway Research Programme Report 89. Factors, trends, and guidelines related to trip length* Highway Research Board (now Transportation Research Board), 2101 Constitution Ave NW, Washington, DC 20418

Bending T, Eden R, 1984 *UK Energy: Structures, Prospects and Policies* (Cambridge University Press, Cambridge)

Berry B J L, 1976, "The counterurbanisation process: urban America since 1970" in *Urbanisation and Counterurbanisation* Urban Affairs Annual Review number 11 (Sage, Beverly Hills, CA) pp 17-30

Borg N, 1981, "Energy usage and activity in the West Midlands 1951-1974" Departmental Publication number 57, Department of Transportation, University of Birmingham, POB 363, Birmingham B15 2TT

Boyden S, Millar S, Newcombe K, O'Neill B, 1981 *The Ecology of a City and its People: The Case of Hong Kong* (Australian National University Press, Canberra)

BRE, 1975, "Energy conservation: a study of energy consumption in buildings and means of saving energy in housing" CP-56, Building Research Establishment, Garston, Watford, Herts

Brooks E, 1976, "On putting the environment in its place: a critique of EIA" in *Environmental Impact Assessment* Eds T O'Riordan, R D Hey, (Saxon House, Farnborough, Hants) pp 167–177

Brown B G H, 1977, "The planning challenge of tidal power" *Town and Country Planning* **48** 159–163

Burchell R W, Listokin D (Eds), 1982 *Energy and Land Use* (Rutgers University, New Brunswick, NJ)

Carpenter S M, Dix M C, 1980, "Perceptions of motoring costs and responses to cost changes" WP-123, Transport Studies Unit, University of Oxford, Oxford

Cherry G, 1982 *The Politics of Town Planning* (Longman, Harlow, Essex)

CHP Group, 1977, "District heating combined with electricity generation in the United Kingdom" *Energy Paper* 20; discussion document prepared by the District Heating Working Party of the Combined Heat and Power Group (HMSO, London)

CHP Group, 1979, "Combined heat and electrical power generation in the United Kingdom" *Energy Paper* 35; report by the Combined Heat and Power Group to the Secretary of State for Energy (HMSO, London)

Christensen B, Jensen-Butler C, 1980, "Energy, planning of heating systems and urban structure" Geographical Institute, University of Århus, 8000 Århus County, Denmark

Christensen B A, Jensen-Butler C, 1982, "Energy and urban structure: heat planning in Denmark" *Progress in Planning* **18** 57–132

City of Davis, 1973, "The Davis General Plan" City of Davis Planning Department, 23 Russell Blvd, Davis, CA 95616

City of Davis, 1982, "Comparative natural gas and kilowatt hour use per capita, Davis, Vacaville, Woodland" City of Davis Planning Department, 23 Russell Blvd, Davis, CA 95616

City of Portland, 1977 *Energy Conservation Choices for the City of Portland* (11 volumes) (US Government Printing Office, Washington, DC)

City of Portland, 1979, "Energy conservation policy" Bureau of Planning, 1120 SW 5th Street, Portland, Oregon

City of Portland, 1982, "Comprehensive land use plan" Bureau of Planning, 1120 SW 5th Street, Portland, Oregon

Clark J W, 1974, "Defining an urban growth strategy which will achieve maximum travel demand reduction and access opportunity enhancement" Research Report 73, 7 UMTA WA 0003, 74 Department of Civil Engineering, Washington University, Seattle, WA 98195

Clifford S, 1979, "EIA—some unanswered questions" *Built Environment Quarterly* **4** 152–160

CCC, 1980, "Clwyd Structure Plan. Submitted written statement" Clwyd County Council Planning Department, Shire Hall, Mold, Clwyd CH7 6NH

Commoner B, 1971, "Power productivity and human welfare" paper presented at the Annual Meeting of the American Association for the Advancement of Science, December; copy available from AAAS, 1515 Massachusetts Ave NW, Washington, DC

Corsi T M, Harvey M E, 1977, "Energy crisis travel behaviour and the transportation planning process" *Transportation Research Record* number 648, pp 30–36

References

Cotgrove S, Duff A, 1980, "Environmentalism, middle class radicalism and politics" *Sociological Review* **28** 333-351

Cotgrove S, Duff A, 1981, "Environmentalism, values and social change" *British Journal of Sociology* **32** 92-110

Courtney M, 1976, "Developing alternative futures" *Journal of the Urban Planning and Development Division, American Society of Civil Engineers* **102** 49-68

CPRE, 1984, "Green belts, housing and conservation" submission to the House of Commons Select Committee on the Environment; Council for the Protection of Rural England, 4 Hobart Place, London SW1W OHY

Craig P P, 1982, "Energy, land use and values; the Davis experience" in *Energy and Land Use* Eds R W Burchell, D Listokin, Centre for Urban Policy Research, Rutgers University, New Brunswick, NJ 08903, pp 510-525

CUEG, 1978, *Seeds for Change* Conservation of Urban Energy Group (Patchwork Press, Melbourne)

CUEG, 1979, "Proposals for energy conservation: Nunawading as a case study" Conservation of Urban Energy Group, Conservation Council of Victoria, 285-287 Little Lonsdale Street, Melbourne, Victoria 3000

Dantzig G D, Saaty T L, 1973 *Compact City: A Plan for a Liveable Urban Environment* (W H Freeman, New York)

Dendrinos D S, 1979, "Energy costs, the transport network, and urban form" *Environment and Planning A* **11** 655-664

Department of Energy, 1976 *Energy Conservation* Cmnd 6575 (HMSO, London)

Department of Energy, 1977, "Report of the Working Group on Energy Elasticities" *Energy Paper* 17 (HMSO, London)

Department of Energy, 1978 *Energy Policy: A Consultative Document* Cmnd 7101 (HMSO, London)

Department of Energy, 1979 *Energy Projections (1979)* Department of Energy, Thames House South, Millbank, London SW1

Department of Energy, 1982 *Proof of Evidence for the Sizewell 'B' Public Inquiry* Department of Energy, Thames House South, Millbank, London SW1

Department of the Environment, 1974, "Structure Plans" circular 98/74 (HMSO, London)

Department of the Environment, 1981, "Local Government, Planning and Land Act 1980. Town and Country Planning: Development Plans" circular 23/81 (HMSO, London)

Department of Transport, 1984 *Buses* Cmnd 9300 (HMSO, London)

Devon CC, 1981 *Devon County Structure Plan* Devon County Council, County Hall, Exeter EX2 4QH

Dickins I, 1975, "Travel patterns and the built environment" *The Planner* **61** 338-340

Dietz T, Vine E, 1982, "Energy impacts of municipal conservation policy" *Energy* **7** 755-758

Dix M C, Goodwin P B, 1981, "Understanding the effects of changing petrol prices: a synthesis of conflicting econometric and psychometric evidence" *Proceedings of the PTRC Annual Meeting* Warwick; copy available from Transport Studies Unit, 11 Bevington Rd, Oxford OX2 6NB

Dix M C, Goodwin P B, 1982, "Petrol prices and car use: a synthesis of conflicting evidence" *Transport Policy Decision Making* **2** 179-195

Doggart J V, 1979, "Eastern flank—energy issues" unpublished discussion paper, Milton Keynes Development Corporation, Wavendon Tower, Wavendon, Milton Keynes MK17 8L

Dorset CC, 1983, "Dorset (excluding South East) Structure Plan, explanatory memorandum" DSP 25, Dorset County Council, Dorset House, 20–22 Christchurch Road, Bournemouth BH1 3NE

Downs A, 1974, "Squeezing spread city" *New York Times Magazine* 17 March, pp 38–47

Eden R, Posner M, Bending R, Crouch E, Stanislaw J, 1981 *Energy Economics: Growth Resources and Policies* (Cambridge University Press, Cambridge)

Edwards J L, Schofer J L, 1975, "Relationships between transportation energy consumption and urban structure: results of simulation studies" Department of Civil and Mineral Engineering, 112 Mines and Metallurgy Building, Minneapolis, MN 55455

Fels M F, Munson M J, 1975, "Energy thrift in urban transportation: options for the future" in *The Energy Conservation Papers: A Report to the Energy Policy Project of the Ford Foundation* Ed. R H Williams (Ballinger, Cambridge, MA) pp 7–104

Fielding A J, 1982, "Counterurbanisation in Western Europe" *Progress in Planning* **17** 1–52

Fleming S C, Short J R, 1984, "Committee rules OK? An examination of planning committee action on officer recommendations" *Environment and Planning A* **16** 965–973

Foley G, Nassim C, 1981 *The Energy Question* 2nd edition (Penguin Books, Harmondsworth, Middx)

Franklin H M, 1974, "Will the new consciousness of energy and environment create an imploding metropolis?" *American Institute of Architects Journal* August issue, pp 28–36

GCC, 1981, "Gloucestershire Structure Plan" Gloucestershire County Council, Shire Hall, Gloucester GL1 2TN

GLC, 1971 *Greater London Transport Survey* (Greater London Council, London)

GLC, 1983 *Draft Alterations to the Greater London Development Plan* (Greater London Council, London)

GLC, 1984 *Summarised Schedule of Responses to Draft Alterations to the Greater London Development Plan* (Greater London Council, London)

GMC, 1979, "Greater Manchester County Structure Plan: submitted written statement" Greater Manchester County Council, County Hall, Piccadilly Gardens, Manchester M60 3HS

GMC, 1981 *Greater Manchester County Structure Plan* Greater Manchester County Council, County Hall, Piccadilly Gardens, Manchester M60 3HS

GMC, 1983, "Energy considerations in the Structure Plan" report of the County Planning Officer, item number 8, 9 February, Greater Manchester County Planning Committee, County Hall, Piccadilly Gardens, Manchester M60 3HS

Granum H, 1976, "Optimisation of thermal insulation in buildings" in *Proceedings of the International CIB Symposium on Energy Conservation in the Built Environment* held at Building Research Station, Watford (Construction Press, Harlow, Essex) pp 371–392

Gray, 1955, "Report on the second national inquiry into domestic expenditure on heating" Building Research Establishment, Garston, Watford, Herts

Grot R A, Socolow R H, 1973, "Energy utilization in a residential community" WP-W-7, School of Engineering, Centre for Environmental Studies, Princeton University, Princeton, NJ 08544

Hafele W, 1981 *Energy in a Finite World* (2 volumes) report by the Energy Systems Program Group of the International Institute for Applied Systems Analysis (Ballinger, Cambridge, MA)

Hall D, 1978, "Energy options and planning" in *Proceedings of the Town and Country Planning Summer School* Royal Town Planning Institute, 26 Portland Place, London W1N 4BE; pp 3-8

Hall D, 1979, "The way ahead: practical proposals for action by planners" in *Proceedings of a Symposium on Energy Policy and Local Planning* Council for the Protection of Rural England, 4 Hobart Place, London SW1W OHY

Hall P, 1974 *Urban and Regional Planning* (Penguin Books, Harmondsworth, Middx)

Hall P (Ed.), 1977 *Europe 2000* (Gerald Duckworth, London)

Halvorsen R, 1974, "Residential demand for electric energy" *Review of Economics and Statistics* **57** 12-18

Halvorsen R, Pollakowski H O, 1979, "Energy prices and housing values" DP-D79-19, Department of City and Regional Planning, Harvard University, Cambridge, MA 02138

Hamer N, 1976 *Getting Nowhere Fast* (Friends of the Earth, London)

Harwood C C, 1977 *Using Land to Save Energy* (The Urban Land Institute, Washington, DC)

Heggie I, 1977, "Consumer response to public transport improvements and car restraint: some practical findings" *Policy and Politics* **5** 47-69

Hemmens G, 1967, "Experiments in urban form and structure" *Highway Research Record* number 207, pp 32-41

Hemphill M L, 1977, "Portland Energy Conservation Project" *Environmental Comment* July issue, pp 14-16

Hillman M, Henderson I, Whalley A, 1973, "Personal mobility and transport policy" PEP Broadsheet 542, Political and Economic Planning, 12 Upper Belgrave Street, London SW1X 8BB

Hills P J, 1983, "Does urban public transport have a future?" in *The Future for the City Centre. Volume 14* Institute of British Geographers Special Publication Series, Eds R L Davies, A G Champion (Academic Press, London) pp 181-202

Howard E, 1898 *Tomorrow: A Peaceful Path to Real Reform* republished in 1945 as *Garden Cities of Tomorrow* (Faber and Faber, London)

IEA, 1981, "Combined heat and power in IEA countries" International Energy Agency, 2 Rue André-Pascal, 75775 Paris Cedex 16

IEA, 1985, "Energy policies and programmes of IEA countries: 1984 review" International Energy Agency, 2 Rue André-Pascal, 75775 Paris Cedex 16

Jamieson G, Mackay W, Latchford J, 1967, "Transportation and land use structures" *Urban Studies* **4** 201-217

Janelle J D, 1969, "Spatial reorganisation: a model and concept" *Annals of the Association of American Geographers* **59** 348-364

Jansson A M, Zucchetto J, 1978, "Man, nature and energy flow on the Island of Gotland" *Ambio* **7** 140-149

Jebson D A, 1981, "The effect of variations in the distribution of heat demand on the cost of district heating networks" paper presented at the District Heating Association's Fourth National Conference, Torquay, June; copy available from Building Research Establishment, Garston, Watford, Herts

JLRC, 1984, "Housing and land 1984-1991: 1992-2000" Joint Land Requirements Committee, The Housing Research Foundation, 58 Portland Place, London W1N 4BU

Kaiser E J, Marsden M E, Burby R J, 1982, "The adoption of energy conservation features in new homes: current practice and proposed policies" in *Energy and Land Use* Eds R W Burchell, D Listokin (Rutgers University Press, New Brunswick, NJ) pp 278-308

Keplinger D, 1978, "Site design and orientation for energy conservation" *Ekistics* **269** 177-180

Keyes D L, 1976, "Energy and land use: an instrument of US conservation policy" *Energy Policy* **4** 108-116

Keyes D L, 1982, "Reducing travel and fuel use through urban planning" in *Energy and Land Use* Eds R W Burchell, D Listokin (Rutgers University Press, New Brunswick, NJ) pp 214-232

Keyes D L, Peterson G, 1977, "Urban development and energy consumption" WP-5049-1.5, The Urban Land Institute, 1200 18th Street NW, Washington, DC

Knowles R L, 1974 *Energy and Form: An Ecological Approach to Urban Growth* (MIT Press, Cambridge, MA)

Leach G, 1976, "Energy futures: wide open to change and choice" *Ambio* **5** 108-116

Leach G, Lewis C, Romig F, van Buren A, Foley G, 1979 *A Low Energy Strategy for the United Kingdom* (Science Reviews, London)

Lee H, 1980, "The role of local governments in promoting energy efficiency" DP-E-80-12, Energy and Environmental Policy Center, John F Kennedy School of Government, Harvard University, Cambridge, MA 02138

Lee H, 1984, "Local energy conservation programmes: past and future" paper written for the Brookings Institution Project on Institutional Barriers to Energy Conservation, Harvard University, Cambridge, MA 02138

Lewis D, 1977, "Estimating the influence of public policy on road traffic levels" *Journal of Transport Economics and Policy* **9** 155-168

Lokmanhekin M, Harvey D G, 1974, "Residential energy consumption: multi family housing. Final report" Hittman Associates, Inc., 9190 Red Branch Road, Columbia, MD 21045

Loudon A, Cornish P, 1975, "Thermal insulation studies" *BRE (Building Research Establishment) News* number 4, page 4

Lovins A, 1977 *Soft Energy Paths* (Penguin Books, Harmondsworth, Middx)

Łucasewiez J, 1978, "Natural resources in Central Europe during the industrial revolution" in *Natural Resources in European History* Eds A Maçzak, W N Parker; conference report, paper R-13, RFF (Resources for the Future), 1755 Massachusetts Avenue, NW, Washington, DC 20036, pp 160-185

McGregor G, 1977, "Davis, California: a pace setting energy conservation city" *Environmental Comment* July issue, pp 16-18

McLoughlin J, 1969 *Urban and Regional Planning: A Systems Approach* (Faber and Faber, London)

Magnan R, Mathieu H, 1975, "Orthopoles, villes en îles" Centre de Recherche d'Urbanisme, 74 Rue de la Federation, 75015 Paris

Maltby D, Monteath I G, Lawler K A, 1978, "The UK surface passenger transport sector: energy consumption and policy options for conservation" *Energy Policy* **6** 294-313

March L, 1967, "Homes beyond the fringe" *RIBA (Royal Institute of British Architects) Journal* August issue, pp 334-337

Marche R, 1980, "Rapport Intermediare" Groupe de Travail 'Demande Voyageurs', Cooperation entre Organismes Nationaux pour l'Etude des Transports Interregionaux (Interim Report of Working Group on 'Travel Demand', Co-operation between national organisations for the Study of Interregional Transport); copy available from Division Transport Interurbaine, Institute de Recherche de Transports, 2 Avenue du Général Malleret-Joinville, 94114 Archueil Cedex

References

Markovitz J, 1971, "Transportation implications of economic cluster development" interim technical report 4245-4424, Tri-State Regional Transportation Commission, New York

Martin L, March L, 1972 *Urban Space and Structures* (Cambridge University Press, Cambridge)

Mathieu H, 1978, "The role of urban planning in relation to overall adaptation to the new energy context: some broad lines of a possible strategic orientation" paper presented at the First International Conference on Energy and Community Development, Athens, July; copy available from Secrétariat du Plan Urbain, 74 Rue de la Federation, 75015 Paris

Meshenberg M J, Ettinger G A, Kron N F, Tschanz J F, 1982, "Guidebook for establishing a local energy management program" Argonne National Laboratory, 9700 South Cass Ave, Argonne, IL 60439

Metcalf A E, 1978, "The perception of car running costs" paper 472/VII/78-EN, Commission of the European Communities, Rue de la Loi 200, B-1049 Brussels

MKDC, 1970, "The plan for Milton Keynes" Milton Keynes Development Corporation, Wavendon Tower, Wavendon, Milton Keynes MK17 8LX

MKDC, 1979, "The east flank—structuring considerations" unpublished internal report, Milton Keynes Development Corporation: Urban Design Unit, Wavendon Tower, Wavendon, Milton Keynes MK17 8LX

MKDC, 1982, Energy Consultative Unit progress report, 1976-1981, "Energy projects in Milton Keynes" Milton Keynes Development Corporation, Wavendon Tower, Wavendon, Milton Keynes MK17 8LX

MKDC, 1983, "Milton Keynes Energy Park: summary development proposal" Milton Keynes Development Corporation, Wavendon Tower, Wavendon, Milton Keynes MK17 8LX

Mogridge M J H, 1977, "An analysis of household transport expenditures 1971-1975" Planning Transportation Research and Computation Annual Summer Meeting, paper G5; copy available from Transport Studies Group, University College London, Gower Street, London WC1N 6BE

Mogridge M J H, 1984, Review of *Future Cities: Spatial Analysis of Energy Issues* by J R Beaumont, P Keys *Progress in Human Geography* **8** 591-593

Murphy T, undated, "CHP/DH systems and the outlook for the inner cities" W S Atkins and Partners, Woodcote Grove, Ashley Road, Epsom, Surrey KT12 5BW

NCC, 1977, "Norfolk Structure Plan: statement on public participation" Norfolk County Council, County Hall, Norwich, Norfolk

Neels K, Cheslow M D, Kirby R F, Peterson G F, 1977, "An empirical investigation of the effects of land use on urban travel" WP-5049-17-1 (The Urban Institute, Washington DC)

Newcombe K, Kalma J D, Aston A R, 1978, "The metabolism of a city: the case of Hong Kong" *Ambio* **7** 3-15

NMA, 1981, "Less energy with more enjoyment in North and West Melbourne" North Melbourne Association, Community Energy Group, PO Box 102, North Melbourne 3051

Ó'Catháin C, Jessop M, 1978, "Density and block spacing for passive solar housing" *Transactions of the Martin Centre for Architectural and Urban Studies* **3** 137-163

Odell P R, 1975 *The Western European Energy Economy—Challenges and Opportunities* The Stamp Memorial Lecture (Athlone Press, London)

Odell P R, 1977, "Energy and planning" *Town and Country Planning* **45** 154-158

Odell P R, 1981, "The energy economy of W Europe: a return to the use of indigenous resources" *Geography* January issue, pp 1-14

Odell P R, 1986, "Draining the world of energy" in *A World in Crisis: Geographical Perspectives on Global Problems* Eds R J Johnston, P J Taylor (Basil Blackwell, Oxford) pp 68-89

Odell P R, Rosing K E, 1983 *The Future of Oil: World Oil Resources and Use* 2nd edition (Kogan Page, London)

Odell P R, Rosing K E, 1984, "The future of oil: a re-evaluation" *OPEC (Organization of Petroleum Exporting Countries) Review* **8** 203-208

Oldfield R H, Bly P H, Webster F V, 1981, "Predicting the use of stage service buses in Great Britain" report LR 1000, Transport and Road Research Laboratory, Crowthorne, Berks

Oregon Land Conservation and Development Commission, 1975, "Statewide planning goals and guidelines" number 13, Department of Land Conservation and Development, 320 SW Stark St, Portland, Oregon

O'Riordan T, 1981 *Environmentalism* 2nd edition (Pion, London)

Owen Carroll T, Udell E B, 1982, "Solar energy, land use and urban form" in *Energy and Land Use* Eds R W Burchell, D Listokin (Rutgers University Press, New Brunswick, NJ) pp 156-177

Owens S E, 1981 *The Energy Implications of Alternative Rural Development Patterns* PhD Thesis, School of Environmental Sciences, University of East Anglia, Norwich NR4 7TJ, Norfolk

Owens S E, 1984a "Energy and spatial structure: a rural example" *Environment and Planning A* **16** 1319-1337

Owens S E, 1984b, "Energy demand and spatial structure" in *Energy Policy and Land Use Planning* Eds D R Cope, P R Hills, P James (Pergamon Press, Oxford) pp 215-240

Owens S E, 1985 "Potential energy planning conflicts in the UK" *Energy Policy* **13** 546-558

Owens S E, 1986, "Strategic planning and energy conservation" *Town Planning Review* **57** 69-86

Owens S E, Rickaby P A, 1983, "Energy and the pattern of human settlements" *Built Environment* **9** 150-159

Pauker G J, 1974, "Can land use management reduce energy consumption for transportation?" paper presented at Caltech Seminar Series *Energy Consumption in Private Transportation,* April; copy available from University of California, Institute of Technology, Pasadena, CA 91125

PCC, 1979, "Structure Plan written statement" Powys County Council, Spa Road East, Llandrindod Wells, Powys LD1 SES

Pindyck R S, 1979 *The Structure of World Energy Demand* (MIT Press, Cambridge, MA)

Platt H L, 1983, "Electric City: regional energy systems and the growth of Chicago" paper presented at the International Seminar on the History of Urban Infrastructure, Ecole Nationale des Ponts et Chaussées, Paris, December; in French in *Les Annales de la Recherche Urbaine* number 23-24, pp 202-217

Poston T, Wilson A G, 1977, "Facility size versus distance travelled: urban services and the fold catastrophe" *Environment and Planning A* **9** 681-686

President's Commission for a National Agenda for the Eighties, 1980 *Urban America in the Eighties: Perspective and Prospects* report of the panel on policies and prospects for metropolitan and nonmetropolitan America (US Government Printing Office, Washington, DC)

RERC, 1974 *The Costs of Sprawl: Environmental and Economic Costs of Alternative Residential Development Patterns at the Urban Fringe* Real Estate Research Corporation (US Government Printing Office, Washington, DC)

References

Rickaby P, 1979, "An energy efficient strategy for the completion of Milton Keynes" Centre for Configurational Studies, Open University, Milton Keynes MK7 6AA

Ridgeway J, Projansky C S, undated *Energy Efficient Community Planning* (J G Press, Emmaus, PA)

Robert S, Randolph G, 1983, "Beyond decentralisation: the evolution of population distribution in England and Wales, 1961–1981" *Geoforum* **14** 75–102

Roberts J S, 1975, "Energy and land use: analysis of alternative development patterns" *Environmental Comment* September issue, pp 2–11

Romanos M C, 1978, "Energy-price effects on metropolitan spatial structure and form" *Environment and Planning A* **10** 93–104

Rose M, 1983, "Urban energy choices: Denver and Kansas City (1900–1940)" paper presented at the International Seminar on the History of Urban Infrastructure, Ecole Nationale des Ponts et Chaussées, Paris, December; in French in *Les Annales de la Recherche Urbaine* number 23–24, pp 181–201

RPA, 1974, "Regional energy consumption" *RPA Bulletin* number 121; Regional Plan Association Inc. and Resources for the Future, 1755 Massachusetts Avenue, NW, Washington, DC 20036

Sassin W, 1981, "Urbanisation and the energy problem" *Options* (International Institute of Applied Systems Analysis news report) **3** 1–4

Schneider J, Beck J, 1973, "Reducing the travel requirements of the American city: an investigation of alternative urban spatial structures" research report 73, US Department of Transportation, Washington, DC

Schumacher D, 1985 *Energy: Crisis or Opportunity?* (Macmillan, London)

Schumacher E F, 1974 *Small is Beautiful; a Study of Economics as if People Mattered* (Sphere Books, London)

Schumacher E F, 1976, "Patterns of human settlement" *Ambio* **5** 91–97

SCE, 1981 *Reports [1980–81]: 1. The Government's Statement on the New Nuclear Power Programme: Volume I* Select Committee on Energy (House of Commons) (HMSO, London)

SCST, 1975 *First Report. Energy Conservation* Select Committee on Science and Technology, *House of Commons Papers–Session 1974–75* **487** (HMSO, London)

Sewell W R D, Foster H D, 1980, "Analysis of the United States experience modifying land use to conserve energy" WP-2, Lands Directorate, Environment Canada, Ministry of Supply and Services, Canada

Smith P, Pollock P, Twiss R, 1978, "Residential solar energy systems: on-site versus district" *Environmental Comment* May issue, pp 4–6

Starkie D, 1982 *The Motorway Age* (Pergamon Press, Oxford)

Steadman P, 1975 *Energy, Environment and Building* (Cambridge University Press, Cambridge)

Steadman P, 1977, "Energy and patterns of land use" *Journal of Architectural Education* **30**(3) 1–7

Steadman P, 1980, "Configurations of land uses, transport networks and their relation to energy use" Centre for Configurational Studies, Open University, Milton Keynes MK7 6AA

Stewart C T Jr, Bennett J T, 1975, "Urban size and structure and private expenditures for gasoline in large cities" *Land Economics* **51** 365–373

Stone P A, 1973 *The Structure, Size and Costs of Urban Settlements* (Cambridge University Press, Cambridge)

The Ecologist 1972, "A blueprint for survival" number 2 (special issue)

Thomas R, Potter S, 1977, "Landscape with pedestrian figures" *Built Environment Quarterly* **3** 286–290

Turrent D, Doggart J, Ferraro R, 1981, "Passive solar housing in the UK" report to the Energy Technology Support Unit, Harwell; copy available from Energy Conscious Design, 11 Emerald Street, London WC1

Van Til J, 1979, "Spatial form and structure in a possible future: some implications of energy shortfall for urban planning" *APA (American Planning Association) Journal* July issue, pp 318–329

Vorhees A M, 1968, "Factors and trends in trip lengths" report 48, National Cooperative Highway Research Program, Washington, DC

WEC, 1980 *World Energy Resources 1985–2020* World Energy Conference (IPC Science and Technology Press, Guildford, Surrey)

Wilbanks T J, 1981, "Local energy initiatives and consensus in energy policy" Energy Division, Oak Ridge National Laboratory, Oak Ridge, Tennessee 37830

Williams W, undated, "The Davis Energy Program" City of Davis, 23 Russell Blvd, Davis, CA 95616

Wilson A G, 1981 *Catastrophe Theory and Bifurcation: Applications to Urban Regional Systems* (Croom Helm, Beckenham, Kent)

Wilson A G, Oulton M J, 1982, "The corner shop to supermarket transition in retailing: the beginnings of empirical evidence" WP-325, School of Geography, University of Leeds, Leeds LS2 9JT

Windheim L S, Wodder R R, 1976, "Cities as energy systems" *Building Systems Design* February/March issue, pp 9–30

WMC, 1980, "West Midlands County Structure Plan" West Midlands County Council, 1 Lancaster Circus, Queensway, Birmingham B4 7DJ

Wood L J, Lee T R, 1980, "Time–space convergence: reappraisal for an oil short future" *Area* **12** 217–222

Wright D, 1979, "Do planners hold the key to energy conservation?" *Planning* number 342, page 6

Wrigley E A, 1962, "The supply of raw materials in the industrial revolution" *Economic History Review* (2nd series) **15** 1–16

Wynne B, 1984, "The institutional context of science, models and policy: the IIASA energy study" *Policy Sciences* **17** 277–320

Zucchetto J, Jansson A-M, 1979, "Integrated regional energy analysis for the island of Gotland, Sweden" *Environment and Planning A* **11** 919–942

Index

Agriculture 2
Århus County, Denmark 101–103
Australia 103–105

Behaviour in response to energy
 constraints 14–18
Building
 layout 4, 22, 46–48
 orientation 4, 22–23, 46–48, 69,
  71, 75, 86, 94
 regulations 71, 85, 86
Built form 21, 41–44, 69, 94

California 22
Catastrophe theory 12
Combined heat and power generation/
 district heating 4, 27, 51–59, 68
 'lead city' schemes 54–56,
  57–58, 65, 75, 94, 99,
  101–103
'Compact city' 32, 62–63, 77
Counterurbanisation 20, 58, 77
Cycling 36, 37, 84, 85, 90, 103

Davis, California 84–87
Decentralisation 11, 20, 59
 of employment and services 33–34,
  64, 75
 of retail facilities 12, 19
Demand elasticities
 residential energy 21
 transport 15–18
Denmark 1, 52, 56, 82, 101–103
Density 4, 10, 22–23, 28, 32–33, 38,
 46, 52–54, 60–61, 69, 99, 102
 and amenity 76–77, 78, 86, 90

Economies of scale 12, 75
Energy budgets 2, 41
Energy conservation 5–8
Energy management programmes
 Milton Keynes 100
 USA 71
Energy prices 3, 5, 6, 10–12, 13, 14,
 15, 19, 21, 54, 75
Energy ratio 6
Environmental impact assessment 72
Environmentalism 90–91
Evaluation in planning process 72

Fuel substitution 21

'Green field' development 38, 67, 79

High-rise development 32, 35, 62, 77
Hobart, Tasmania 17
Housing market 21–22, 43–44

Impedance (in gravity model) 14, 19,
 31, 34
Infilling 38, 48
Inner cities 10, 13
 combined heat and power
  generation in 57–58

Land uses
 clustering of 35, 38
 interspersion of 28, 33–35, 38, 60,
  77, 103, 106
Linear development 60, 90
Linear grid 38, 61, 65–67, 77, 97
Local government 82
Local planning authorities 8–9, 29,
 83–84
London 29, 31, 33, 43
 Greater London Development Plan
  92, 93–95

Melbourne 103–105
Microclimate 22, 44, 94, 96
Middle East 6
Milton Keynes 67, 83, 97–101
 Energy Park 99–101, 105

Neighbourhood concept 64
New York 33, 35, 43
Norfolk, United Kingdom 34–35
Nuclear power 12, 82
Nucleated urban form 63–65, 77
Nunawading, Australia 105

Pedestrian facilities 36
Portland, Oregon 70, 87–91
Public participation 79, 82, 86, 90,
 102

Randers, Denmark 101
Recreation 64
Renewable energy (*see also* solar energy)
 2, 4, 12, 27, 35, 61, 65, 68, 94, 99
 and spatial structure 44–52
 'soft energy paths' 49–51
Residential energy consumption
 21–33, 85
Retail facilities 12, 19
 'out-of-town' 79

Scandinavia 1, 52, 80
Sheffield
    bus fares policy 39
Solar acres 22
Solar energy
    active 48–49
    and district heating 58
    passive 22, 46–49, 84, 96–97
    and solar electricity 48–49
Space heating 3, 11, 21, 27, 41, 48, 68, 90, 101
Structure plans 36, 70, 77, 80, 83–84, 91–93
    Bedfordshire 77
    Berkshire 79
    Cheshire 74
    Clwyd 83
    Cornwall 77
    Devon 39
    Dorset 92
    East Sussex 92
    Greater Manchester 83, 92
    Leicestershire 83
    Merseyside 92
    Norfolk 79
    Nottinghamshire 77
    Powys 92
    South Yorkshire 77
    Tyne and Wear 77
    West Midlands 92
    West Sussex 34
Suburbanisation 10, 20, 58
    of employment 20
Suburbs 4, 13, 20, 37, 38, 49, 67

Thermal insulation 42, 69, 84, 97
Time-space connectivity 20
Time-space convergence 11, 20
Time-space divergence 20
Traffic management 36, 38
Transport 2, 3, 4, 10, 27, 68, 94, 99, 103
    energy conservation in 28–40, 61, 68, 89–90
    energy efficiency of different modes of 36–37
    household expenditure on 16
    response to fuel shortages in 15–18
    rural 39
Transport policy 36–40, 79, 83
Tyneside 'Metro' 39

United Kingdom 1, 2, 3, 8, 12, 20, 22, 27, 29–30, 33, 38–40, 41, 43, 48, 52–56, 69, 71, 77, 80, 82, 91–101
Urban models 18–21, 25–27, 30, 33, 67
Urban shape 32, 38
Urban size 29–32, 60
Urban sprawl 3, 10, 31, 48, 85
Urban trends 10–18
Utility analysis 19

Vehicle efficiency 69
Vehicle restraint 39–40

Zoning 90